The Impact of LED (Light-Emitting Diode) Spectra on Plant Growth and Quality

The Impact of LED (Light-Emitting Diode) Spectra on Plant Growth and Quality

Editors

Hail Rihan
Mick Fuller

MDPI

Basel • Beijing • Wuhan • Barcelona • Belgrade • Novi Sad • Cluj • Manchester

Editors

Hail Rihan	Mick Fuller
School of Biological and	School of Biological and
Marine Sciences	Marine Sciences
University of Plymouth	University of Plymouth
Plymouth	Plymouth
United Kingdom	United Kingdom

Editorial Office
MDPI
St. Alban-Anlage 66
4052 Basel, Switzerland

This is a reprint of articles from the Special Issue published online in the open access journal *Plants* (ISSN 2223-7747) (available at: www.mdpi.com/journal/plants/special_issues/led_spectra_growth).

For citation purposes, cite each article independently as indicated on the article page online and as indicated below:

Lastname, A.A.; Lastname, B.B. Article Title. *Journal Name* **Year**, *Volume Number*, Page Range.

ISBN 978-3-0365-8641-0 (Hbk)
ISBN 978-3-0365-8640-3 (PDF)
doi.org/10.3390/books978-3-0365-8640-3

Contents

About the Editors

Hail Rihan

Hail Rihan, PhD, MRSB, FHEA, is a Professor of Plant Biotechnology and a Senior Research Fellow in the School of Biological and Marine Sciences at the University of Plymouth, UK. He is also a Consultant Professor at King Abdulaziz City for Science and Technology (KACST), KSA, and a Visiting Professor at the University of Jeddah, KSA. Prof. Rihan is a globally recognised expert in the rapidly growing field of controlled-environment agriculture (CEA) and plant tissue culture. He has led public–private collaborations in the field of CEA and published over 55 peer-reviewed papers, making several key contributions to the field of plant science in the past 15 years. His research work focuses on the commercial application of next-generation LED technologies for intensive hydroponic culture of high-value crops.

Mick Fuller

Michael Fuller, PhD, is a Professor Emeritus of Plant Physiology in the School of Biological and Marine Sciences at the University of Plymouth, UK. Prof. Fuller has a distinguished career as a plant scientist and senior researcher. His research interests include plant stress biology, plant physiology, plant biotechnology, agronomy, and especially controlled-environment agriculture (CEA). He has held many research and applied research grants and published over 200 papers and reports. He was Head of the Graduate School at the University of Plymouth for 12 years, Chair of the UK Council for Graduate Education, and a member of the Steering Council for Graduate Education for the European University Association. He had supervised over 50 postgraduate research students.

Preface

The revolution in controlled-environment agriculture (CEA) has been driven by developments in light-emitting diodes (LEDs) and their value engineering to make these technologies affordable for use in many plant science applications. A key feature of LEDs is their narrow waveband of light production, allowing the development of arrays of LEDs to match the spectral absorption characteristics of plants. This has necessitated intense research activities to define and characterize the best spectral combinations and light intensities to be used for various crop species. This Special Issue presents several pioneering articles showing early key advances in this field of research.

Hail Rihan and Mick Fuller
Editors

plants

MDPI

Article

The Impact of Light Spectrum and Intensity on the Growth, Physiology, and Antioxidant Activity of Lettuce (*Lactuca sativa* L.)

Shiren J. Mohamed [1,2,*], Hail Z. Rihan [1,3], Naofel Aljafer [1] and Michael P. Fuller [1]

[1] School of Biological and Marine Sciences, University of Plymouth, Plymouth PL4 8AA, UK; hail.rihan@plymouth.ac.uk (H.Z.R.); naofel.aljafer@plymouth.ac.uk (N.A.); mfuller@plymouth.ac.uk (M.P.F.)
[2] Biotechnology and Crop Sciences Department, College of Agricultural Engineering Sciences, University of Sulaimani, Sulaimani 46001, Iraq
[3] Phytome Life Sciences, Launceston PL15 7AB, UK
* Correspondence: shiren.mohamed@univsul.edu.iq

Abstract: This study focused on the physiology, growth and antioxidant activity response of hydroponically grown lettuce (*Lactuca sativa* L.) under sole-source LED lighting of differing spectra. Lighting spectra were provided by differing combinations of LEDs of three different peak wavelengths, (Blue 435, Blue 450, and Red 663 nm) with ratios of B450/R663: 1.25 ± 0.1, B450/R663: 1.25 ± 0.1, and B450/R663 1:1 at two light intensities of photosynthetically active radiation (PAR) (270 μmol m^{-2} s^{-1} and 60 μmol m^{-2} s^{-1}). A further experiment was conducted, in which Blue and Red LEDs were supplemented with Green (Blue 450, Red 663, and Green 520 nm) with ratios of B435/R663: 1.25 ± 0.1, B450/R663/G520: 1/0.73/0.26, and B450/R663: 1.25 ± 0.1. LED light intensities under the different spectra were adjusted to deliver the same level of PAR (270 ± 20 μmol m^{-2} s^{-1}). Results from the first experiment showed that increased fraction of blue 435 nm in combination with red light at 663 nm at high irradiance enhanced the physiology of lettuce (i.e., significantly increased assimilation rate, stomatal conductance and transpiration rate) and increased the yield while having no significant effect on antioxidant activity. At the lower irradiance, the B435/R663 significantly increased antioxidant activity compared to other spectra. Results from the second experiment showed no significant effect of the spectra of LEDs on the physiology and yield of lettuce, but antioxidant activity was very significantly induced by B450/R663 at the ratio of 1.25 ± 0.1. However, the amount was still less than that obtained by B435/R663 1.25 ± 0.1 from the first experiment. This study indicates that LED light with a spectrum of B435/R663 at a ratio of 1.25 ± 0.1 significantly improves lettuce yield and antioxidant activity.

Keywords: LEDs; lettuce; physiology; fresh weight; antioxidant activity

Citation: Mohamed, S.J.; Rihan, H.Z.; Aljafer, N.; Fuller, M.P. The Impact of Light Spectrum and Intensity on the Growth, Physiology, and Antioxidant Activity of Lettuce (*Lactuca sativa* L.). *Plants* **2021**, *10*, 2162. https://doi.org/10.3390/plants10102162

Academic Editor: Masayuki Fujita

Received: 22 September 2021
Accepted: 30 September 2021
Published: 12 October 2021

1. Introduction

Lettuce (*Lactuca sativa* L.) belongs to the family Composite, and is an important dietary leafy salad vegetable that is primarily consumed fresh or in salad mixes [1]. It is a major source of bioactive compounds with diverse biological activities: it has antioxidant, anti-inflammatory, anticancer, antimicrobial, cholesterol lowering, and antidiabetic effects, and it is a good source of fibre, iron, folate, and vitamin C [2–5]. Lettuce is widely grown in semi-controlled environments in glasshouses and plastic tunnels, often using hydroponic culture [6,7]. Some lettuce crops are now grown under controlled conditions using artificial light in plant factories. Lettuce is frequently used as a test species when investigating the optimisation of plant factory conditions.

Light is one of the fundamental environmental factors for plant growth and development. Light quality, in comparison with light intensity and photoperiod, has been shown to have a much more complex impact on plant physiology and morphology in terms of spectral distribution, since specific wavelengths stimulate different physiological and morphological responses [8] This indicates the importance of chlorophyll A&B. Chlorophyll

A is the primary pigment of photosynthesis and absorbs light from 430 nm to 662 nm. Chlorophyll A has a central role in the transference of energy to the reaction centre and contributes very significantly to the electron transport chain, since it donates two excited electrons. Chlorophyll B absorbs a blue light range between 453 nm to 642 nm, chlorophyll B helps organisms to convert the energy from light spectra to chemical energy. Furthermore, chlorophyll B can absorb a wider range of wavelengths of light, which enables more energy to be transferred to chlorophyll A [9].

Light-emitting diodes (LEDs) have higher luminous efficiency, long life, and higher efficacy, leading to reduced associated heating [10] compared to other artificial lighting sources, such as fluorescent bulbs or sodium vapour lamps. Furthermore, in indoor plant factory farming systems, LEDs allow for the modification of the spectrum to fit the plant species requirements. Lettuce is widely cultivated in plant factories under LEDs [11,12], because of its adaptability to controlled environments, its short growth cycle, and defined rosette shoot shape [13]. Plant Factories (controlled environment agriculture) are new forms of agriculture that are not dependent on arable land and that can be developed in the urban environment are gaining increasing popularity. Plant Factories with Artificial Lighting (PFALs) or Vertical Farms with Artificial Lighting (VFALs), are closed plant production systems where environmental factors (e.g., temperature, humidity, light, and CO_2 concentration) are controlled, minimizing the interactions with the external climate. There is a significant growing interest in this form of farming because it can be built anywhere, high resource use efficiency (water, CO_2, fertilizer, etc.) can be achieved with minimum emission of pollutants to the outside environment, the growing environment is not affected by the outside climate and soil fertility, production can be year-round and productivity is over 100 times that of field production, produce quality such as concentrations of phytonutrients can be enhanced through manipulation of the growing environment, especially light quality; produce is pesticide-free and need not be washed before eating; produce has a longer shelf life because the bacterial load is generally less than 300 CFU g^{-1}, which is 1/100 to 1/1000 that of field-grown produce and energy for transportation can be reduced by building PF near urban areas. Light is a key factor and a very important element for the Plant Factories since it has direct impact of growth, yield, and quality of plants.

It has been reported that lettuce grown under combined Red and Blue LEDs exhibit the highest chlorophyll content, photosynthesis rate [14], pigment content, leaf numbers, leaf area index and shoot dry weight, and also increased antioxidant activity [15]. However, plants under monochromatic Blue or Red LEDs have displayed growth abnormality and reduced photosynthetic rate [14,16]. Recently, Naznin et al. [15] reported that lettuce grown solely under Red LEDs had significantly reduced biomass, chlorophyll content, carotenoid content, and antioxidant levels. Moreover, it has been observed that the lettuce plants could not perform normally in Red light only, and the combination of 90% Red and 10% Blue was considered more effective [17]. Photosynthesis rate, stomatal density, growth, and mineral element content under a combination of Red and Blue appears to be dependent on the Red light/Blue light ratio (R/B ratio and all these parameters increased with a decrease in R/B ratio) [14]. Pennisi et al. [10] reported that, when the R/B ratio increased from 0.5 to 3, the chlorophyll and flavonoid content, nutrient uptake and water use efficiency of the lettuce leaves improved, with a resultant yield increase of 1.6-fold, although no further increase was reported when the R/B exceeded a ratio of 3. It has been reported that the optimal ratio of R/B for lettuce is at an intensity of 200 μmol m^{-2} s^{-1} irradiance for 16 h for highest photosynthesis rate and stomatal conductance is R/B = 1 compared to ratios of R/B of 4, 8, 12 with a significant decrease when the ratio of R to B increased from 1 to 12 [14].

The synergistic effectiveness of the combined Red to Blue ratio can be more clear on lettuce growth in term of leaf area and dry weight when a small quantity of green G light (24%) is added, since green light is better able to penetrate the plant canopy than Red or Blue light [18]. This may be because the plants have sensitive green light sensors (phytochromes and cryptochrome), although their efficiency in processing green is less than that shown in response to blue and red wavelengths [19]. In contrast, Saito et al. [20]

reported that lettuce plants under monochromic Red light had a higher photosynthetic rate, greater leaf number and greater fresh weight compared to either blue light or a mixture of RB light. These findings are supported by Wang et al. [14], who concluded that Red light might be the most effective wavelength for photosynthesis and growth in lettuce (*Lactuca sativa* L.). Lee and Kim [21] also concluded that Red light LEDs with a peak of 634 nm and 659 nm and Blue light LEDs with a peak of 450 nm are the potential spectral wavelengths that boost the photosynthetic rate most effectively, leading to increased leaf area, shoot fresh weight, leaf chlorophyll, and anthocyanin content.

However, there is less agreement regarding optimal fraction of either R or B combination effect on lettuce. The current study therefore aimed to investigate the different fraction of R and B LEDs and different RB ratios on the physiology, growth and antioxidant activities in lettuce (*Lactuca sativa* L.).

2. Materials and Methods

2.1. Plant Material and Growth Condition

Lettuce seeds were obtained from CN seeds (CN Seeds Ltd., Pymoor, UK), then sown and germinated in the greenhouse at Skardon Gardens. When seedlings had their first pair of true leaves, they were transferred to the plant factory facility at the University of Plymouth. The university's plant factory facility is a converted insulated greenhouse where external light has been excluded. The multi-tier hydroponic growing system consists of gulleys for NFT (nutrient film technique) and is installed with interchangeable LED light units. The plant factory is divided into several multi-shelf hydroponic units, each consisting of three tiers. The distance between tiers is 50 cm, and 16 plants were planted in each tier at a spacing of 20 cm within a gully and 20 cm between gullies. The temperature and humidity were monitored, using Gemini data loggers (Tinytag Plus (part No GP-1590)) and an instantaneous thermometer (Fisher Scientific, Hampton, NH, USA) at 23 ± 2 °C. The light/dark period was set to 16/8 h.

Two experiments were established:

2.1.1. First Experiment

Three lighting treatments were designed and applied at two intensities (high: 270 and low 60 μmol m^{-2} s^{-1}) measured using a UPRtek MK350N premium Standalone handheld spectral light meter, Taiwan. Light treatments were as follows:

Blue 435 nm rich treatment: Blue/Red (B435/R663): Blue rich spectrum with 435 nm wavelength used as a source of blue (B/R: 1.25 ± 0.1) (Blue 435 nm to Red 663 nm spectrum peak ratio, 1.6:1).

Blue 450 nm rich treatment: Blue/Red B450/R663: Blue rich spectrum with 450 nm wavelength used as a source of blue (B/R: 1.25 ± 0.1) (Blue 450 nm to Red 663 peak ratio, 1.6:1).

Red rich treatment: Blue/Red treatment (B/R-rich): Red 663 nm rich light spectrum with 450 nm wavelength used as a source of blue (B/R: 0.72) Blue to Red ratio, (1:1). (Figures 1 and 2).

2.1.2. Second Experiment

Three lighting treatments at 170 ± 10 μmol m^{-2} s^{-1} were designed as follows:

(1) Blue 450 nm rich treatment: Blue/Red treatment (B-rich/R). Blue rich spectrum with 450 nm wavelength used as a source of blue (B/R: 1.25 ± 0.1) (Blue (450 nm) to Red (663) peak ratio, 1.6:1).
(2) Blue, red, green treatment: Blue/Red/Green treatment (B/R/G). Blue rich spectrum with 450 nm wavelength used as a source of blue with (B/R/G: 1.25/1/0.35) (Blue (435 nm) to Red (663) peak ratio, 1.6:1).
(3) Red rich treatment: Red (663 nm) rich light spectrum with 450 nm wavelength used as a source of blue (B/R: 0.72) (Blue Red peak ratio 1:1.2).

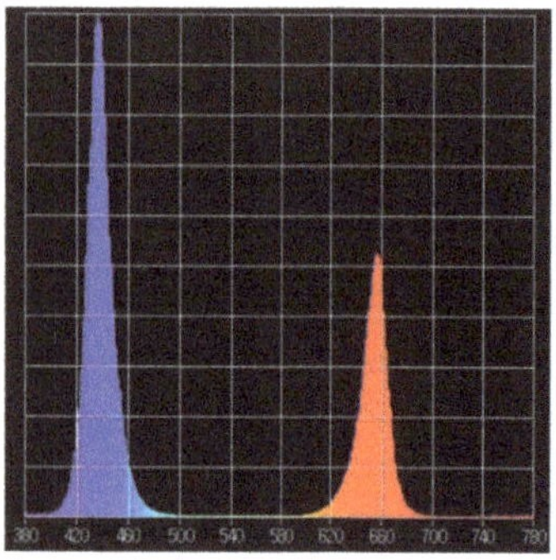

(**A**) Blue 435 nm rich

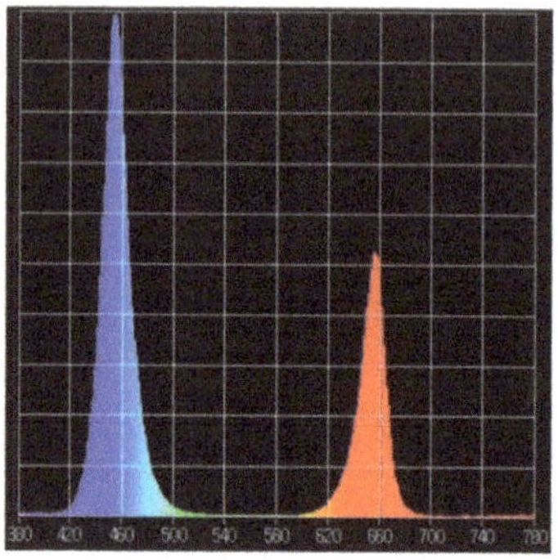

(**A**) Blue 450 nm rich

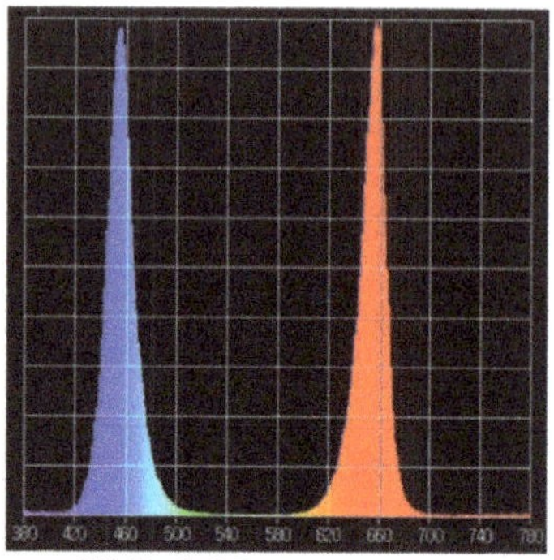

(**A**) Red rich treatment

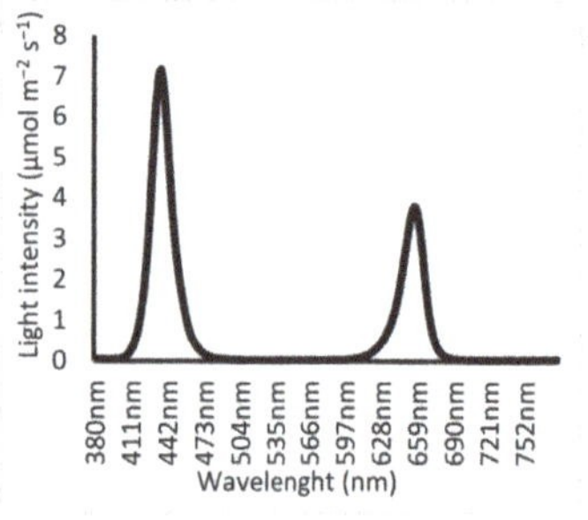

(**B**) Blue 435 nm rich

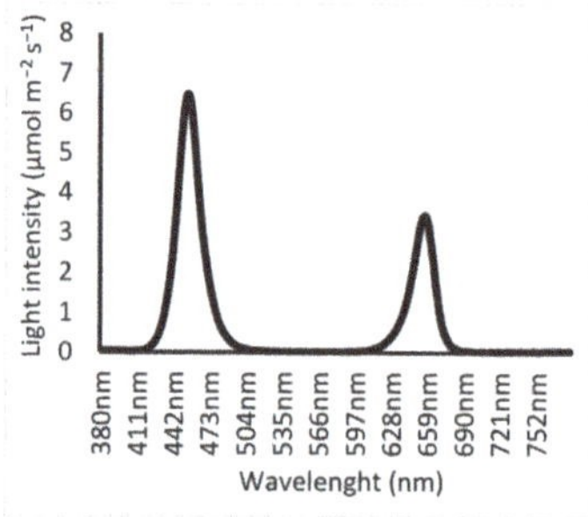

(**B**) Blue 450 nm rich

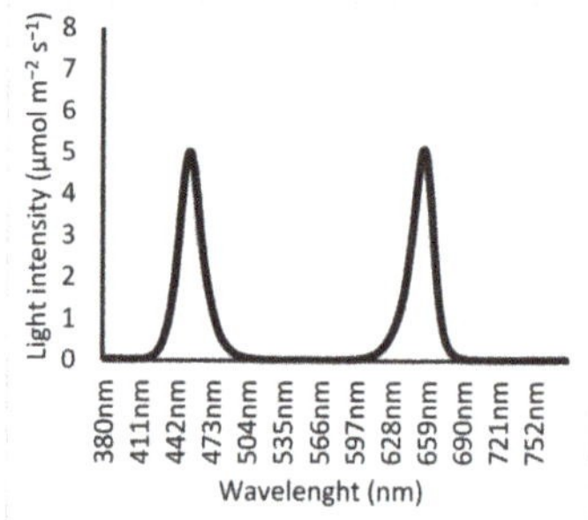

(**B**) Red rich treatment

Figure 1. Spectra of the LED treatments (Blue 435 nm rich treatment, Blue 450 nm rich treatment and Red rich treatment), as measured by an UPRtek spectrophotometer: (**A**) the relative light intensity. (**B**) The radiant density of the light spectrum intensity.

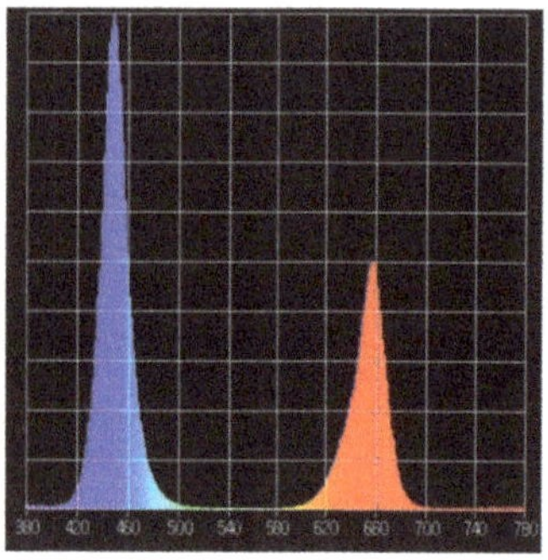

(**A**) Blue 450 nm rich treatment

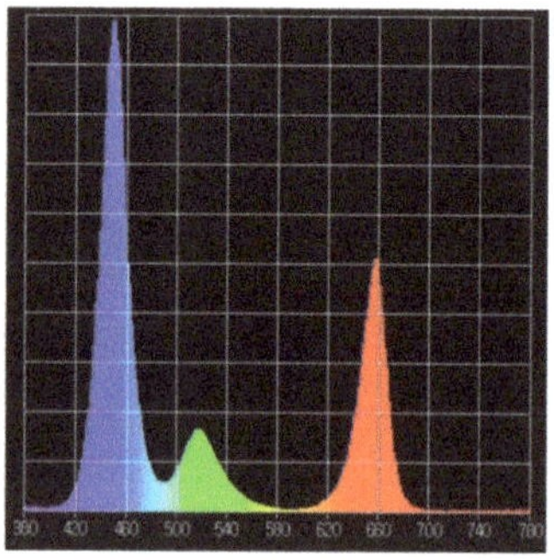

(**A**) Blue, red, green treatment

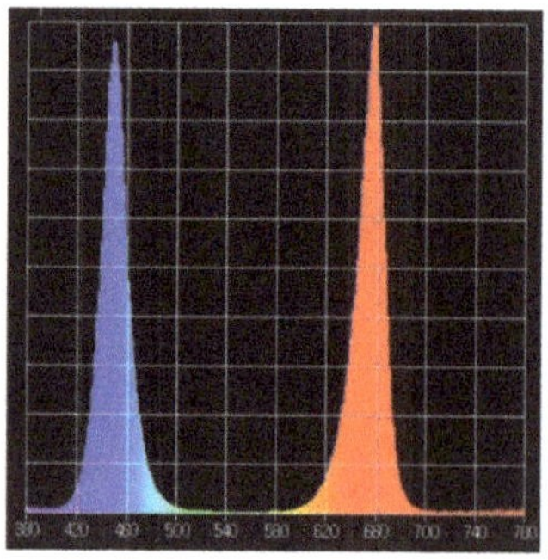

(**A**) Red rich treatment

Figure 2. *Cont.*

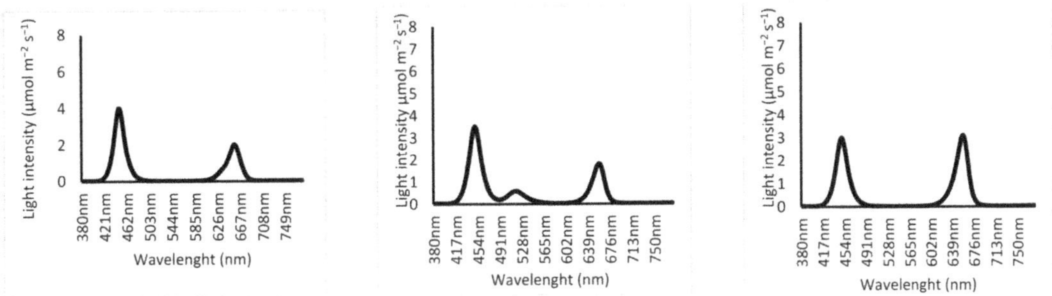

(**B**) Blue 450 nm rich treatment (**B**) Blue, red, green treatment (**B**) Red rich treatment

Figure 2. Spectra of the LED treatments (Blue/Red treatment (B-rich/R), blue, red, green treatment and red rich treatment) as measured by an UPRtek spectrophotometer: (**A**) the relative light intensity. (**B**) The radiant density of the light spectrum intensity.

2.2. Physiological Parameters Measurements

Physiological response (assimilation rate µmol m^{-2} s^{-1}, stomatal conductance mmol m^{-2} s^{-1}, and transpiration rate mmol m^{-2} s^{-1}) of planted lettuce to the lighting treatments was measured at two stages of development: at the initial vegetative, stage 4 weeks from the transplanting of plantlets; and the second (final) harvest stage, conducted 7 weeks after transplanting plantlets to the plant factory setting. The three unfolded top leaves were chosen from five plants from each treatments. Physiological measurements included light-saturated instantaneous maximum photosynthetic rate Amax (µg cm^{-2} s^{-1}) was measured using an LCi-SD Highly Portable Ambient Photosynthesis System (ADC BioScientific, Herts, UK).

2.3. Determination of Plant Morphology

Morphological response of planted lettuce to the lighting treatments were measured at two stages of development: at the initial vegetative, stage 4 weeks from the transplanting of plantlets; and the second (final) harvest stage, conducted 7 weeks after transplanting plantlets to the plant factory setting. Morphological measurements of five randomly chosen plants from each treatment were taken. These included leaf number (cmshoot fresh weight (FW); and root fresh weight (RFW) (g), using a sensitive Fisher Scientific SG-402 laboratory balance.

2.4. Antioxidant Activity Analysis

The plants (all plants) from the second cut were stored in a deep freezer (at −20 °C) and freeze-dried for antioxidant analysis. The total antioxidant activity was analysed using the Ferric Reducing Ability of Plasma (FRAP) assay [22]. The method is based on the reduction of Fe^{3+} TPTZ complex (colourless complex) to Fe^{2+}-tripyridyltriazine (blue coloured complex) formed by the action of electrons donating antioxidants at low pH. This reaction was monitored by measuring the change in absorbance at 595 nm. The Ferric reducing antioxidant power (FRAP) reagent was prepared by mixing 10 part of 300 mM acetate buffer, 1 part of 10 mL TPTZ in 40 mM HCl and 1 part 20 mM $FeCl_3 \cdot 6H_2O$. For the extraction, 0.1 g of freeze dried leaves were weighed and ground using a mortar and homogenized with 4 mL of HEPES buffer and sand purified by acid. From this solution, 0.80 mL was placed in an Eppendorf tube and centrifuged at 13,000 rpm in a microfuge (Micro star 12) for 2 min. The extract was then stored on ice prior to use. The calibration curve was prepared by plotting the absorbance at 595 nm versus different concentrations (0, 0.2, 0.4, 0.6, 0.8 and 1 MM) of $FeSO_4$. The concentrations of $FeSO_4$ were in turn plotted against a concentration of standard antioxidant Trolox. The blank was prepared by mixing the 0.80 FRAP with 0.10 µL HEPES buffer and the spectrophotometer (Bibby Scientific Ltd., Stone, UK) set to zero against the blank. From the each of the stored extract samples,

0.10 μM of each stored extract were transferred to the cuvettes and 0.80 mL of FRAP reagent added. The FRAP values were obtained by comparing the absorbance at 595 change in the test mixture with those obtained from increasing concentrations of Fe^{3+} and expressed as mg of Trolox equivalent per gram of sample.

2.5. Statistical Analysis

All data were subjected to analysis of variance (ANOVA) using Minitab software (version 17), and comparisons of means were made using the least significant difference (LSD) test at a 5% level of probability.

3. Results

3.1. Assimilation Rate, Stomatal Conductance and Transpiration at High Lighting Intensity

Assimilation rate in lettuce cultivated under all LED treatments was reduced when the plants reached maturity, (Table 1), which is the assimilation rate at first harvest is significantly ($p = 0$) higher than at second harvest. At both harvest stages, the photosynthesis rate showed a remarkable ($p = 0.003$) difference between LED treatments. At the first harvest, the B435/R significantly increased assimilation rate by 26% compared to the B450/R, while there were no significant differences between B435/R and B/R-rich. At second harvest, there were significant differences between all LED treatments. The B435/R significantly increased the assimilation rate by about 100 and 32% in comparison to B450/R and B/R-rich, respectively. As with the assimilation rate, stomatal conductance in plant leaves grown under all LED treatments significantly ($p = 0.00$) decreased with plants' maturity (Table 1) and was significantly lower at second harvest compared to first harvest. There was a significant ($p = 0.00$) difference in leaves stomatal conductance between LED treatments at both harvest stages. At the first harvest, the greatest stomatal conductance was under B/R-rich, which increased stomatal conductance by 150 and 400% compared to B435/R and B450/R, respectively. At the second harvest, the greatest stomatal conductance was under B435/R, which was greater by 95 and 290% compared to B450/R and B/R-rich. In contrast to the assimilation rate and stomatal conductance, the transpiration rate increased with plant maturity, (Table 1). Significant ($p = 0.066$) differences between harvest stages were observed, and there was a significant ($p = 0.004$) effect of LED treatments on transpiration rate at both harvest stages. At the first harvest stage, the highest transpiration rate was under B/R-rich, which increased by about 33 and 100% compared to B435/R and B450/R, respectively. At the second harvest stage, the highest value was at B/R-rich, followed by B435/R and then B450/R.

3.2. Growth and Morphology at High Lighting Intensity

As shown in Table 1, all LED treatments stimulated lettuce plant growth. Plants produced greater fresh weight (shoot and root) and leaf numbers at second harvest compared to first harvest. At both harvest stages, plants produced different fresh weights (shoot + root) under different LED treatments. Shoot fresh weight (Table 1) was significantly ($p = 0.003$) increased when plants were grown under combination of B435/R, compared to plants grown under B450/R and B/R-rich. At second harvest, the B435/R increased the plants' fresh weight by 36 and 14% compared to B450/R and B/R-rich, respectively, and the B/R-rich increased plant fresh weight by 13% as compared to B450/R. Furthermore, the highest root fresh weight was produced by plants cultivated under B435/R, compared to other treatments (Table 1). At the second harvest, the B435/R significantly ($p = 0083$) increased the root fresh weight by about 50 and 46% as compared to B450/R and B/R-rich, respectively, and, similarly to shoot fresh weight, the B/R-rich increased the root fresh weight by about 25% compared to B450/R. With regard to the number of leaves, no significant effect between all LED treatments were observed. (Table 1).

Table 1. The effects of high light intensity of different LED treatments on A: Assimilation rate (for harvest stage ($p \leq 0.001$), for light treatment ($p = 0.003$) for interaction between light treatments and harvest stage ($p = 0.605$)). B: Stomatal conductance (for harvest stage ($p \leq 0.001$), for light treatments ($p \leq 0.001$) and for interaction between light treatments and harvest stage ($p \leq 0.001$)). C: transpiration rate LSD for harvest stage ($p = 0.06$) for treatments ($p = 0.066$), and for interaction light treatments and harvest stage ($p = 0.001$). D: Shoot fresh weight (g) (for harvest stages ($p \leq 0.001$), for treatment ($p = 0.003$) and for interaction between harvest stage and light treatments ($p = 0.08$)). E: Root fresh weight (g) for harvest ($p \leq 0.001$) for light treatments ($p = 0.083$) and interaction ($p = 0.31$). F: Leaves number (for harvest stage ($p \leq 0.001$), for light treatment ($p = 0.199$) and for interaction between harvest stage and light treatments ($p = 0.153$)).

Light Treatment	Harvest Stage	Growth and Physiological Parameters					
		Assimilation Rate ($\mu mol\ m^{-2}\ s^{-1}$)	Stomatal Conductance ($mmol\ m^{-2}\ s^{-1}$)	Transpiration Rate ($mmol\ m^{-2}\ s^{-1}$)	Shoot Fresh Weight (g)	Leaves Number	Root Fresh Weight (g)
Blue/Red (B435/R663)	Harvest stage 1	6.58 ± 0.31	0.5 ± 0.150	1.19 ± 0.021	84.9 ± 6.84	39.37 ± 4.30	8.6 ± 0.67
	Harvest stage 2	3.94 ± 0.33	0.38 ± 0.054	1.54 ± 0.106	165.30 ± 2.44	69.67 ± 6.09	11.62 ± 1
Blue/Red (B450/R663)	Harvest stage 1	5.03 ± 0.43	0.23 ± 0.024	1.20 ± 0.066	77.36 ± 8.83	25.25 ± 3.03	3.96 ± 0.87
	Harvest stage 2	1.95 ± 0.31	0.12 ± 0.012	1.23 ± 0.092	119.22 ± 2.45	67.83 ± 6.24	7.61 ± 0.64
Blue/Red treatment (B/R-rich)	Harvest stage 1	6.90 ± 0.47	1.09 ± 0.162	1.48 ± 0.074	60.31 ± 11.98	24.5 ± 4.52	4.22 ± 0.37
	Harvest stage 2	2.76 ± 0.36	0.20 ± 0.017	2.05 ± 0.126	137.01 ± 2.44	71.67 ± 8.86	8.83 ± 1.19
LSD	Harvest stage	2.41	0.40	Not significant	40.49	10.17	4.39
	Light treatment	1.96	0.33	1.25	33.06	Not significant	Not significant
	Interaction between light treatment and harvest stage	Not significant	0.23	0.70	Not significant	Not significant	Not significant

3.3. *Assimilation Rate, Stomatal Conductance and Transpiration at Low Lighting Intensity*

All physiological parameters (assimilation rate, stomatal conductance, and transpiration rate) for plants cultivated by all LED treatments with lower light intensity progressively decreased with plant maturity (Table 2), as indicated by a significant reduction of these parameters at second harvest, compared to first harvest. At both harvests, there were significant effects of LED treatment on physiological parameters. Assimilation rate was significantly stimulated ($p = 0.003$) under B 435/R compared to B450/R, with no significant differences between B345/R and B/R-rich, which both increased assimilation rate by 25% compared to B450/R at first harvest. However, at the second harvest, the B435/R increased assimilation rate by 135 and 53% compared to B450/R and B/R-rich, respectively, whereas, at the first harvest, the highest ($p = 0.072$) leaf stomatal conductance was at B450/R, with an increment by 67% compared to other LED treatments. At the second harvest, the highest value of stomatal conductance was obtained by B435/R and B/R-rich, with an increment of 100% compared to B450/R. Although the transpiration rate significantly differed ($p = 0.003$) between LED treatments, the greatest value at the first harvest was under B435/R, with an increment by 60 and 33% as compared to B450/R and B/R-rich, respectively. Furthermore at the second harvest, the greatest value for the transpiration rate was at B345/R, with an increment of about 114% compared with B450/R, and no significant differences between B435/R and B/R–rich were record.

3.4. *Growth and Morphology at Low Lighting Intensity*

In general, plants cultivated by all LED lighting progressively increased growth in term of plant fresh weight (shoot + root), as in Table 2, which shows the shoot weight very significantly greater ($p \geq 0.001$) and root fresh weight significantly ($p = 0$) greater at second harvest than at first harvest. Plants cultivated at both B435/R and B450/R produced significant ($p = 0.078$) shoot fresh weight (Table 2) which increased shoot fresh weight by 50% compared to B/R-rich at second harvest. Whereas the highest root fresh weight obtained by B 435/R first, then followed by B/R-rich compared to B 450/R (Table 2). Which was at second harvest, the B 350/R increased root fresh weight by about 49 and 22% compared to B450/R and B/R-rich, respectively. There were no significant differences in the number of lettuce leaves grown under different LED treatments (Table 2).

3.5. *Second Experiment*

3.5.1. Assimilation Rate, Stomatal Conductance and Transpiration under B-rich/R, B-rich/R/G and B/R-rich

As in the first experiment, all the physiological parameters; assimilation rate, stomatal conductance, and transpiration rate for plants cultivated by all LEDs progressively decreased when plants reached maturity (Table 3). At both harvest stages, there were no significant ($p = 0.062$) differences between all B-rich/R, B-rich/R/G and B/R-rich lights on the plants' assimilation rates. Stomatal conductance significantly ($p = 0.002$) has a clear difference between all LEDs treatments, at first harvest B/R-rich increased the stomatal conductance by (20 and 67%) as compared to B-rich/R and B-rich/R/G, respectively. Similarly, the transpiration rate showed significant ($p = 0.00$) differences between LED treatments. The highest values were obtained under B/R-rich, which increased transpiration rate by 94 and 280%, compared to B-rich/R and B-rich/R/G, respectively, at first harvest.

Table 2. The effects of low light intensity of different LEDs treatments on A: Assimilation rate (for harvest stage ($p \leq 0.001$), for light treatments ($p = 0.003$), and for interaction between harvest stage and light treatments ($p = 0.605$)). B: Stomatal conductance (for harvest stage ($p \leq 0.001$), for treatment ($p = 0.072$), and for interaction between harvest stage and light treatments ($p \leq 0.001$)). C: transpiration rate (for harvest ($p = 0.002$) for light treatments ($p = 0.003$) and for interaction between harvest stage and light treatments ($p = 0.31$)). D: Shoot fresh weight (g) (for harvest stages ($p = 0.001$), for light treatments ($p = 0.078$), and interaction between harvest stage and light treatments ($p = 0.069$)). E: Root fresh weight (g) (for harvest ($p \leq 0.001$) for light treatments ($p = 0.013$), and for interaction between harvest stage and light treatments ($p = 0.210$)), F: leaves number (for harvest stage ($p \leq 0.001$), for light treatments ($p = 0.07$) for interaction between harvest stage and light treatments ($p = 0.012$)).

Light Treatment	Harvest Stage	Growth and Physiological Parameters					
		Assimilation Rate (µmol m^{-2} s^{-1})	Stomatal Conductance (mmol m^{-2} s^{-1})	Transpiration Rate (mmol m^{-2} s^{-1})	Shoot Fresh Weight (g)	Leaves Number	Root Fresh Weight (g)
Blue/Red (B435/R663)	Harvest stage 1	3.18 ± 0.64	0.16 ± 0.016	3.83 ± 0.81	29.6 ± 2.85	23.25 ± 2.37	7.57 ± 1.19
	Harvest stage 2	2.68 ± 0.62	0.09 ± 0.011	1.45 ± 0.1	75.87 ± 2.45	40.83 ± 3.81	10.18 ± 0.81
Blue/Red (B450/R663)	Harvest stage 1	2.44 ± 0.31	0.26 ± 0.019	2.38 ± 0.10	22.69 ± 3.27	21.5 ± 1.90	1.87 ± 0.22
	Harvest stage 2	1.04 ± 0.11	0.07 ± 0.007	0.93 ± 0.07	73.99 ± 2.45	46 ± 2.84	7.58 ± 0.61
Blue/Red treatment (B/R-rich)	Harvest stage 1	3.12 ± 0.37	0.16 ± 0.016	3.01 ± 0.69	26.62 ± 2.55	24.12 ± 2.41	3.79 ± 0.36
	Harvest stage 2	1.85 ± 0.30	0.11 ± 0.009	1.46 ± 0.10	44.98 ± 2.45	28.83 ± 3.92	9.05 ± 0.9
LSD	Harvest stage	2.41	0.073	0.54	30.19	5.06	2.55
	Light treatment	1.96	Not significant	0.44	Not significant	Not significant	2.08
	Interaction between light treatment and harvest stage	Not significant	0.04	0.31	Not significant	13.01	Not significant

Table 3. The effects of low light intensity of different LEDs treatments on A: Assimilation rate (LSD for harvest stages = 2.26 ($p = 0.00$), for treatments = no significance ($p = 0.062$), and for interaction = 1.31 ($p = 0.00$)). B: Stomatal conductance (LSD for harvest stages = no significance ($p = 0.590$), for treatments = 0.09 ($p = 0.002$), and for interaction = no significance ($p = 0.118$)). C: transpiration rate (LSD for harvest = 1.42 ($p = 0.038$), for treatments = 1.16 ($p = 0.00$), and for interaction = 0.82 ($p = 0.010$)). D: Shoot fresh weight (g) (LSD for harvest = 43.05 ($p \leq 0.001$), for treatments = no significance ($p = 0.598$), and for interaction = no significance ($p = 0.723$)). E: Root fresh weight (g) (LSD for harvest = 3.35 ($p = 00$), for treatments = no significance ($p = 0.86$), and for interaction = no significance ($p = 0.072$)). F: Leaves number (for harvest stage ($p \leq 0.001$), for light treatment ($p = 0.277$) for the interaction between harvest stage and light treatments ($p = 0.044$)).

Light Treatment	Harvest Stage	Growth and Physiological Parameters					
		Assimilation Rate ($\mu mol\ m^{-2}\ s^{-1}$)	**Stomatal Conductance ($mmol\ m^{-2}\ s^{-1}$)**	**Transpiration Rate ($mmol\ m^{-2}\ s^{-1}$)**	**Shoot Fresh Weight (g)**	**Leaves Number**	**Root Fresh Weight (g)**
Blue/Red (B435/R663)	Harvest stage 1	3.30 ± 0.74	0.15 ± 0.01	1.58 ± 0.16	37.72 ± 2.72	21.62 ± 1.13	5.24 ± 1.04
	Harvest stage 2	2.76 ± 0.36	0.21 ± 0.017	2.05 ± 0.126	100.5 ± 2.45	46.5 ± 5.46	10.67 ± 0.57
Blue/Red (B450/R663)	Harvest stage 1	4.04 ± 0.37	0.19 ± 0.027	1.20 ± 0.066	33.81 ± 3.36	23 ± 2.84	1.89 ± 0.14
	Harvest stage 2	2.27 ± 0.27	0.15 ± 0.016	1.41 ± 0.085	95.33 ± 2.45	50.33 ± 4.66	6.04 ± 0.43
Blue/Red treatment (B/R-rich)	Harvest stage 1	6.39 ± 0.45	0.25 ± 0.010	3.84 ± 0.729	24.87 ± 4.79	24.75 ± 1.77	2.26 ± 0.51
	Harvest stage 2	2.05 ± 0.20	0.22 ± 0.020	1.89 ± 0.091	82.21 ± 2.45	36 ± 2.73	10.23 ± 1.04
LSD	Harvest stage	2.26	Not significant	1.42	43.05	5.30	3.35
	Light treatment	Not significant	0.09	1.16	Not significant	Not significant	Not significant
	Interaction between light treatment and harvest stage	1.31	Not significant	0.82	Not significant	14.71	Not significant

3.5.2. Growth and Morphology under B-rich/R, B-rich/R/G and B/R-rich

Plants grown under all lighting treatment showed an increase in terms of shoots and root fresh weight when plants reached maturity. However, there were no significant ($p = 0.598$) differences between LEDs treatments on plant fresh weight (shoot + root fresh weight) Table 3. Furthermore no significant differences were recorded for plant leaf numbers grown under different LED treatments (Table 3).

3.6. *Antioxidant Activity*

At second harvest, there were no significant ($p = 0.296$) differences between all LED treatments with high intensity (270 ± 20 μmol m^{-2} s^{-1}). Lettuce shoot antioxidant activity (in μmol $FeSO_4$ L^{-1}) was recorded, as shown in Figure 3A. The same LED treatments showed a highly significant ($p = 0$) difference in lettuce antioxidant activity (in μmol $FeSO_4$ L^{-1}) when the intensity of lights became lower 60 μmol m^{-2} s^{-1} (Figure 3B), and the highest level of antioxidant activity (in μmol $FeSO_4$ L^{-1}) was at B345/R, followed by B450/R, with increasing rate by about 53 and 420%, compared with B450/R and B/R-rich, respectively. In addition, in the second experiment, as shown in Figure 3C, there were significant ($p = 0.004$) differences between LED treatment in lettuce shoot antioxidant activity levels (in μmol $FeSO_4$ L^{-1}), the highest level was at B-rich/R, with an increment of 125 and 260%, as compared with B-rich/R/G and B/R-rich, respectively.

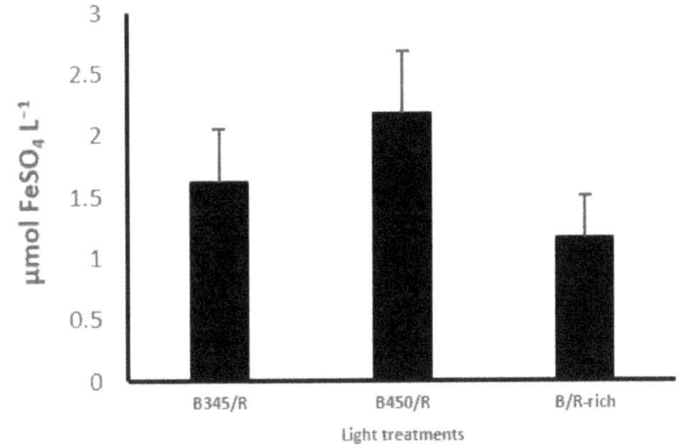

(**a**)

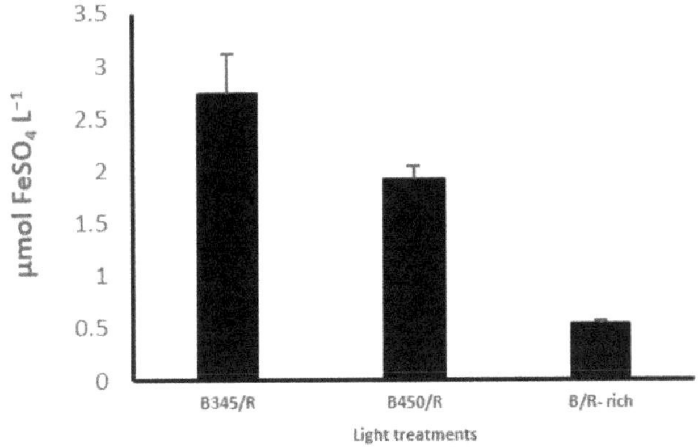

(**b**)

Figure 3. *Cont.*

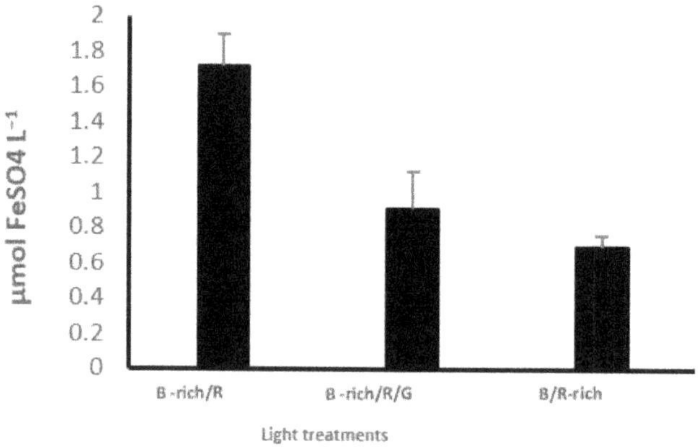

(**c**)

Figure 3. Antioxidant activities in lettuce at second harvest under different LEDs treatments; (**a**) effects of high intensity of B345/R, B450/R and B/R-rich: LSD of treatments = no significance (p = 0.296). (**b**) Effects of low intensity of B345/R, B450/Rand B/R-rich: LSD treatments = 0.60 (p = 0). (**c**) Effects of B-rich/R, B-rich/R/G and B/R-rich; LSD treatments 0.41 (p = 0.004).

4. Discussion

The results from the first experiment at the higher intensity (270 ± 20 µmol m^{-2} s^{-1}) confirmed that, firstly: using blue light with a wavelength peak of 435 nm enhanced assimilation rate by 26 and 100% at first and second harvest, respectively, and this was also observed at the lower intensity (60 µmol m^{-2} s^{-1}). The B435/R enhanced assimilation rate by 25 and 135% for the first and second harvest, respectively, more than blue light with a peak of 450 nm. This finding supports our previous study on sweet basil (*Ocimum basilicum*) [23,24]. The second confirmation was that the use of blue light with a wavelength peak of 450 nm could match the absorbance of lettuce pigments when the ratio of B450/R is (1:1) in the balance of LEDs in array increased assimilation rate. In the current study, the stomatal conductance and transpiration rate at the first harvest did not reach the greatest value at B435/R compared to B450/R. At the second harvest, the B435/R gave the highest value of stomatal conductance and transpiration rate with an increment by 95 and 100% of stomatal conductance and by 435 and 114% of transpiration rate, compared to B450/R for first and second intensities, respectively. These findings indicate that the blue light with a wavelength of 435 nm gradually enhanced stomatal conductance, until reaching the highest level at the second harvest.

The third confirmation n was that for lettuce shoot fresh weight, using blue light with wavelength peaks of 435 nm and 450 nm at low light intensity had same effect as that for B435/R and B450/R treatments. The same results for lettuce have been recently reported by [14]: they found that lettuce assimilation rate and fresh weight increased with decreasing R/B ratio from (12 to 1) and this was also associated with an increase in stomatal conductance. This result is in agreement with that of Yan et al. [7], who reported that the assimilation rate in lettuce leaves increased with decrease in R/B ratio. This was due the inhibition of photorespiration and stimulation of stomatal opening to CO_2 uptake for assimilation [25]. Similarly, increasing in the red light fraction decreased stomatal conductance in lettuce [14] due to the fact that the guard cells of stomata were opened by the blue light phototropin receptors [26]. As a result, plants under blue LEDs maintained photosynthesis more effectively than under red LEDs [27]. Blue (B) and red (R) wavelengths of light are absorbed more by photosynthetic pigments than by other wavelengths. A wide range of plant physiology and growth processes, such as leaf photosynthesis function, photo-morphogenesis, phototropism, stomatal opening [28], hypocotyl elongation, leaf expansion, leaf anatomy, enzyme synthesis, gene expression, and chloroplast movement are driven by blue light [29,30]. On the other hand, red light causes stem elongation and increases chlorophyll content and photosynthesis [31].

Our results are consistent with those of Pennisi et al. [10] who showed that a ratio of R/B = 3 with light intensity of 215 μmol m^{-2} s^{-1} increased chlorophyll content and decreased photosystem II quantum efficiency and transpiration rate, resulting in increased water use efficiency and a maximised lettuce yield. A higher R/B ratio did not result in additional lettuce yield. However, [32] observed that assimilation rate in lettuce decreased as blue light fraction increased from 20% to 30%. The explanation for this is that red light is the most efficient wavelength for photosynthesis. As reported by McCree, (1971), the relative quantum efficiency of red light (600–700 nm) was higher than that of blue light (400–500 nm) because blue fraction was absorbed by flavonoids in vacuoles and/or anthocyanin's pigments and is less efficient in transferring energy to the reaction centres for photosynthesis [33].

In general, it has been measured at 90% of red and blue light LEDs absorbance by plant [34]. This finding indicates that both red and blue strongly affect plant physiology and development. Plants grown under red LEDs exhibited photosynthesis and growth similar to those grown under blue LEDs [35].

In the current research, the root fresh weight at the low light intensity of 60 μmol m^{-2} s^{-1} under B435/R was higher by 49 and 22% compared to B450/R and B/R-rich, respectively, while both B435/R and B450/R produced the same amount of shoot fresh weight. The reason for this is that increased ("ratio of"?) blue light with wavelength peak of 435 nm to red light with a wavelength peak 663 nm and at a low intensity of LED light can alter the assimilation translocation between lettuce plant organs. A similar tendency was found in lettuce when the ratio of blue to red increased by 20–50% [17].

At both intensities, there was no significant difference in the number of lettuce leaves among plants cultured under all LEDs treatments. In contrast, the combination of R/B LEDs significantly increased leaf numbers in lettuce [14,35–37].

In the second experiment, Table 3 shows non-significant differences between all LED treatments B-rich/R (1.6:1), B/R/G (1/0.73/0.26) and B/R-rich (1:1.2) on assimilation rate, while stomatal conductance and transpiration rate significantly increased by B/R-rich, compared to B-rich/R and B/R/G, respectively. This indicated that the supplementation of a small amount of green light to red and blue light could achieve maximum assimilation rate, similar to a combination of red and blue [38]. Moreover, green light can be absorbed by cytochrome (cry), decreasing the activity of chromophores on cry and leading to the induction of stomata opens on leaves by blue light that is absorbed by cry [39]. On the other hand, altering the ratio of blue to red light could be sufficient for lettuce assimilation rate. It is known that the wavelengths of both red and blue lights are necessary in the process of plant photosynthesis [40], indicating that light quantities are more effective than light quality in lettuce production, as recently reported by [7]. As a consequence, the lettuce fresh shoot biomass, root fresh biomass, and leaf numbers did not significantly differ in all LED treatments. These results contradict the findings of Shao et al. [41] who observed that shoot fresh weight in lettuce increased by 20.5% under RBG LEDs with light intensity of 150 μmol m^{-1} s^{-1}. This was the result of an increase in the assimilation rate of 24.2% compared to R/B LEDs, because green light is able to penetrate the plant canopy and supply energy, especially in plants with overlapping leaves, such as lettuce. Furthermore, [18] suggested that red and blue LEDs with 24% green treatment gave the highest shoot fresh weight and plants under RBG LEDs and RB had similar assimilation rates. In addition, [42,43] considered the RBG LEDs as an optimal combination wavelength for lettuce growth [44].

In the present study, the levels of antioxidant activity levels in lettuce shoots (in μmol $FeSO_4$ L^{-1}) under B435/R (1.25 ± 0.1) compared to B450/R (1.25 ± 0.1) B/R-rich (1:1) at the first experiment with low light intensity of 60 μmol m^{-2} s^{-1} but not at the high intensity of (270 ± 20 μmol m^{-2} s^{-1}). Moreover, at B/R-rich (1.25 ± 0.1) compared to B/R/G (1/0.73/0.26) and B435/R (1:1.2) at second experiment (Figure 3) were significantly higher than other LED treatments. This was in agreement with Son et al. [45] who reported that increasing the fraction of blue combined with red increased the concentration of phenolic acid and antioxidant activity by 41% at R/B (6:4) and by 24% at R/B (8:2), compared to R/B (9:1) with light intensity of (130 ± 7 μmol m^{-2} s^{-1} 130 ± 7 μmol m^{-2} s^{-1}) 12 h photoperiod. The blue light is effective in accumulating secondary metabolism and in promoting the phenolic concentration and antioxidant capacity resulting from an activation of the PAL gateway enzyme in the biosynthesis of phenolic, enhanced by monochromatic blue LEDs [46]. It has frequently been reported that increasing the blue light during light period induced the concentration of many bioactive compounds: this effect has been reported in several cultivars of lettuce [47,48]. More recently, Naznin et al. [15] demonstrated that a higher proportion of blue light to red (83% R + 17% B) compared to (91% R + 9% B and 95% R + 5% B) with intensity of ±200 μmol m^{-2} s^{-1} increased antioxidant activity in lettuce leaves. However, using combined blue and red light led to less change in secondary metabolism and it has been suggested that the metabolic process is more sensitive to change in monochromatic light [45], and the metabolic changes in either monochromatic light or combined light may be induced by differences in the activation of photoreceptors, such as phytochromes and cryptochromes effectively absorbing blue and red light [49].

5. Conclusions

The results of this study showed that the growth, photosynthesis, and antioxidant activity of lettuce performed better with a combination of blue light with a peak wavelength of 435 nm and red with a peak wavelength of 663 nm with a ratio of (1.25 ± 0.1), than with a combination of blue light with a peak wavelength of 450 nm and red with a peak wavelength of 663 nm with a ratio of (1.25 ± 0.1) at high intensity of (270 ± 20 μmol m^{-2} s^{-1}). However, when a small amount of green light with a wavelength peak at 520 nm is added to the combination of B450/R663 nm and the ratio of B450 nm to Red663 nm is the same or R663/B450 nm = 1.2, all of the LEDs enhanced the assimilation rate by the same amount and produced the same amount of lettuce fresh weight. From these results, it can be concluded that B435/R at high intensity is the best LED for the production of economic yields of hydroponically grown lettuce in the plant factory for production of the highest yields. It was also found that B435/R at the low intensity of 60 μmol m^{-2} s^{-1} is the best LED for producing the highest level of antioxidant activity.

Grow LED light system is a growing area for both research and commercial applications. This will increase the capacity of testing more specific wavelengths in both red and blue regions of lights. This is currently is still one of the limitations for a deeper understanding of the plant response to light spectrum.

Author Contributions: S.J.M. Investigation, laboratory analysis & writing the paper; H.Z.R. experiment establishment and data analysis; N.A. Laboratory analysis; M.P.F. Writing review & editing. All authors have read and agreed to the published version of the manuscript.

Funding: This research was funded by Kurdistan Ministry of Higher Education and University of Sulaimani/College of Agricultural Engineering Sciences for the financial support to carry out this research. This research was also partially funded by the Seale-Hayne Educational Trust (grant RR204307).

Institutional Review Board Statement: Not applicable.

Informed Consent Statement: Not applicable.

Conflicts of Interest: The authors declare no conflict of interest.

References

1. Chon, S.-U.; Jang, H.-G.; Kim, D.-K.; Kim, Y.-M.; Boo, H.-O. Allelopathic potential in lettuce (*Lactuca sativa* L.) plants. *Sci. Hortic.* **2005**, *106*, 309–317. [CrossRef]
2. Briskin, D.P. Medicinal Plants and Phytomedicines. Linking Plant Biochemistry and Physiology to Human Health. *Plant Physiol.* **2000**, *124*, 507–514. [CrossRef]
3. Havsteen, B.H. The biochemistry and medical significance of the flavonoids. *Pharmacol. Ther.* **2002**, *96*, 67–202. [PubMed]

4. Teodoro, A.J. Bioactive Compounds of Food: Their Role in the Prevention and Treatment of Diseases. *Oxidative Med. Cell. Longev.* **2019**, *2019*, 1–4. [CrossRef] [PubMed]
5. Tsao, R.; Deng, Z. Separation procedures for naturally occurring antioxidant phytochemicals. *J. Chromatogr. B* **2004**, *812*, 85–99. [CrossRef]
6. Lee, J.G.; Lee, B.Y.; Lee, H.J. Accumulation of phytotoxic organic acids in reused nutrient solution during hydroponic cultivation of lettuce (*Lactuca sativa* L.). *Sci. Hortic.* **2006**, *110*, 119–128. [CrossRef]
7. Yan, Z.; He, D.; Niu, G.; Zhou, Q.; Qu, Y. Growth, Nutritional Quality, and Energy Use Efficiency of Hydroponic Lettuce as Influenced by Daily Light Integrals Exposed to White versus White Plus Red Light-emitting Diodes. *HortScience* **2019**, *54*, 1737–1744. [CrossRef]
8. Amoozgar, A.; Mohammadi, A.; Sabzalian, M.R. Impact of light-emitting diode irradiation on photosynthesis, phytochemical composition and mineral element content of lettuce cv. Grizzly. *Photosynthetica* **2016**, *55*, 85–95. [CrossRef]
9. Shams, M.; Ekinci, M.; Turan, M.; Dursun, A.; Kul, R.; Yildirim, E. Growth, nutrient uptake and enzyme activity response of Lettuce (*Lactuca sativa* L.) to excess copper. *Environ. Sustain.* **2019**, *2*, 67–73. [CrossRef]
10. Pennisi, G.; Orsini, F.; Blasioli, S.; Cellini, A.; Crepaldi, A.; Braschi, I.; Spinelli, F.; Nicola, S.; Fernandez, J.A.; Stanghellini, C. Resource use efficiency of indoor lettuce (*Lactuca sativa* L.) cultivation as affected by red: Blue ratio provided by LED lighting. *Sci. Rep.* **2019**, *9*, 1–11. [CrossRef]
11. Shimizu, H.; Saito, Y.; Nakashima, H.; Miyasaka, J.; Ohdoi, K. Light Environment Optimization for Lettuce Growth in Plant Factory. *IFAC Proc. Vol.* **2011**, *44*, 605–609. [CrossRef]
12. Saito, K.; Ishigami, Y.; Goto, E. Evaluation of the Light Environment of a Plant Factory with Artificial Light by Using an Optical Simulation. *Agronomy* **2020**, *10*, 1663. [CrossRef]
13. Goins, G.D.; Ruffe, L.M.; Cranston, N.A.; Yorio, N.C.; Wheeler, R.M.; Sager, J.C. Salad Crop Production under Different Wavelengths of Red Light-Emitting Diodes (LEDs). SAE Technical Paper (Canada). 2001. Available online: https://www.sae.org/publications/technical-papers/content/2001-01-2422/ (accessed on 29 September 2021).
14. Wang, J.; Lu, W.; Tong, Y.; Yang, Q. Leaf Morphology, Photosynthetic Performance, Chlorophyll Fluorescence, Stomatal Development of Lettuce (*Lactuca sativa* L.) Exposed to Different Ratios of Red Light to Blue Light. *Front. Plant Sci.* **2016**, *7*, 250. [CrossRef]
15. Naznin, M.T.; Lefsrud, M.; Gravel, V.; Azad, O.K. Blue Light added with Red LEDs Enhance Growth Characteristics, Pigments Content, and Antioxidant Capacity in Lettuce, Spinach, Kale, Basil, and Sweet Pepper in a Controlled Environment. *Plants* **2019**, *8*, 93. [CrossRef]
16. Wang, X.Y.; Xu, X.M.; Cui, J. The importance of blue light for leaf area expansion, development of photosynthetic apparatus, and chloroplast ultrastructure of Cucumis sativus grown under weak light. *Photosynthetica* **2014**, *53*, 213–222. [CrossRef]
17. Tang, Y.K.; Guo, S.S.; Ai, W.D.; Qin, L.F. Effects of Red and Blue Light Emitting Diodes (LEDs) on the Growth and Development of Lettuce (var. Youmaicai). SAE Technical Paper (Canda). 2009. Available online: https://www.sae.org/publications/technical-papers/content/2009-01-2565/ (accessed on 29 September 2021).
18. Kim, H.-H.; Goins, G.D.; Wheeler, R.M.; Sager, J.C. Green-light Supplementation for Enhanced Lettuce Growth under Red- and Blue-light-emitting Diodes. *HortScience* **2004**, *39*, 1617–1622. [CrossRef]
19. Folta, K.M.; Maruhnich, S.A. Green light: A signal to slow down or stop. *J. Exp. Bot.* **2007**, *58*, 3099–3111. [CrossRef]
20. Saito, Y.; Shimizu, H.; Nakashima, H.; Miyasaka, J.; Ohdoi, K. The effect of light quality on growth of lettuce. *IFAC Proc. Vol.* **2010**, *43*, 294–298. [CrossRef]
21. Lee, J.S.; Kim, Y.H. Growth and Anthocyanins of Lettuce Grown under Red or Blue Light-emitting Diodes with Distinct Peak Wavelength. *Korean J. Hortic. Sci. Technol.* **2014**, *32*, 330–339. [CrossRef]
22. Benzie, I.; Strain, J. The Ferric Reducing Ability of Plasma (FRAP) as a Measure of "Antioxidant Power": The FRAP Assay. *Anal. Biochem.* **1996**, *239*, 70–76. [CrossRef] [PubMed]
23. Rihan, H.Z.; Aldarkazali, M.; Mohamed, S.J.; McMulkin, N.B.; Jbara, M.H.; Fuller, M.P. A Novel New Light Recipe Significantly Increases the Growth and Yield of Sweet Basil (*Ocimum basilicum*) Grown in Plant Factory System. *Agronomy* **2020**, *10*, 934. [CrossRef]
24. Landi, M.; Zivcak, M.; Sytar, O.; Brestic, M.; Allakhverdiev, S.I. Plasticity of photosynthetic processes and the accumulation of secondary metabolites in plants in response to monochromatic light environments: A review. *Biochim. Biophys. Acta (BBA) Bioenerg.* **2020**, *1861*, 148131. [CrossRef] [PubMed]
25. Chen, X.-L.; Xue, X.-Z.; Guo, W.-Z.; Wang, L.-C.; Qiao, X.-J. Growth and nutritional properties of lettuce affected by mixed irradiation of white and supplemental light provided by light-emitting diode. *Sci. Hortic.* **2016**, *200*, 111–118. [CrossRef]
26. Hiyama, A.; Takemiya, A.; Munemasa, S.; Okuma, E.; Sugiyama, N.; Tada, Y.; Murata, Y.; Shimazaki, K.-I. Blue light and CO_2 signals converge to regulate light-induced stomatal opening. *Nat. Commun.* **2017**, *8*, 1–13. [CrossRef] [PubMed]
27. Savvides, A.; Fanourakis, D.; van Ieperen, W. Co-ordination of hydraulic and stomatal conductances across light qualities in cucumber leaves. *J. Exp. Bot.* **2011**, *63*, 1135–1143. [CrossRef]
28. Whitelam, C.G.; Halliday, K.J. *Annual Plant Reviews, Light and Plant Development*; John Wiley & Sons: Hoboken, NJ, USA, 2008; Volume 30.
29. Inoue, S.-I.; Kinoshita, T.; Matsumoto, M.; Nakayama, K.I.; Doi, M.; Shimazaki, K.-I. Blue light-induced autophosphorylation of phototropin is a primary step for signaling. *Proc. Natl. Acad. Sci. USA* **2008**, *105*, 5626–5631. [CrossRef] [PubMed]

30. Wang, H.; Gu, M.; Cui, J.; Shi, K.; Zhou, Y.; Yu, J. Effects of light quality on CO_2 assimilation, chlorophyll-fluorescence quenching, expression of Calvin cycle genes and carbohydrate accumulation in Cucumis sativus. *J. Photochem. Photobiol. B Biol.* **2009**, *96*, 30–37. [CrossRef] [PubMed]
31. Kang, J.H.; KrishnaKumar, S.; Atulba, S.L.S.; Jeong, B.R.; Hwang, S.J. Light intensity and photoperiod influence the growth and development of hydroponically grown leaf lettuce in a closed-type plant factory system. *Hortic. Environ. Biotechnol.* **2013**, *54*, 501–509. [CrossRef]
32. Kang, W.H.; Park, J.S.; Park, K.S.; Son, J.E. Leaf photosynthetic rate, growth, and morphology of lettuce under different fractions of red, blue, and green light from light-emitting diodes (LEDs). *Hortic. Environ. Biotechnol.* **2016**, *57*, 573–579. [CrossRef]
33. Zivcak, M.; Brückova, K.; Sytar, O.; Brestic, M.; Olsovska, K.; Allakhverdiev, S.I. Lettuce flavonoids screening and phe-notyping by chlorophyll fluorescence excitation ratio. *Planta* **2017**, *245*, 1215–1229. [CrossRef]
34. Terashima, I.; Fujita, T.; Inoue, T.; Chow, W.S.; Oguchi, R. Green light drives leaf photosynthesis more efficiently than red light in strong white light: Revisiting the en-igmatic question of why leaves are green. *Plant Cell Physiol.* **2009**, *50*, 684–697. [CrossRef]
35. Olle, M.; Viršilė, A. The effects of light-emitting diode lighting on greenhouse plant growth and quality. *Agric. Food Sci.* **2013**, *22*, 223–234. [CrossRef]
36. Ohtake, N.; Ishikura, M.; Suzuki, H.; Yamori, W.; Goto, E. Continuous Irradiation with Alternating Red and Blue Light Enhances Plant Growth While Keeping Nutritional Quality in Lettuce. *HortScience* **2018**, *53*, 1804–1809. [CrossRef]
37. Shin, Y.-S.; Lee, M.-J.; Lee, E.-S.; Ahn, J.-H.; Kim, M.-K.; Lee, J.-E.; Do, H.-W.; Cheung, J.-D.; Park, J.-U.; Um, Y.-G.; et al. Effect of Light Emitting Diodes Treatment on Growth and Quality of Lettuce (*Lactuca sativa* L. 'Oak Leaf'). *J. Life Sci.* **2014**, *24*, 148–153. [CrossRef]
38. Sytar, O.; Zivcak, M.; Neugart, S.; Toutounchi, P.M.; Brestic, M. Precultivation of young seedlings under different color shades modifies the accumulation of phenolic compounds in Cichorium leaves in later growth phases. *Environ. Exp. Bot.* **2019**, *165*, 30–38. [CrossRef]
39. Sellaro, R.; Crepy, M.; Trupkin, S.A.; Karayekov, E.; Buchovsky, A.S.; Rossi, C.; Casal, J.J. Cryptochrome as a Sensor of the Blue/Green Ratio of Natural Radiation in Arabidopsis. *Plant Physiol.* **2010**, *154*, 401–409. [CrossRef]
40. Taiz, L.; Zeiger, E.; Møller, I.M.; Murphy, A. *Plant Physiology and Development*; Sinauer Associates Incorporated: Sunderland, MA, USA, 2015.
41. Shao, M.; Liu, W.; Zha, L.; Zhou, C.; Zhang, Y.; Li, B. Differential effects of high light duration on growth, nutritional quality, and oxidative stress of hydroponic lettuce under red and blue LED irradiation. *Sci. Hortic.* **2020**, *268*, 109366. [CrossRef]
42. Li, L.; Tong, Y.X.; Lu, J.L.; Li, Y.M.; Yang, Q.C. Lettuce growth, nutritional quality, and energy use efficiency as affected by red–blue light combined with different monochromatic wavelengths. *HortScience* **2020**, *55*, 613–620. [CrossRef]
43. Zhang, T.; Maruhnich, S.A.; Folta, K.M. Green Light Induces Shade Avoidance Symptoms. *Plant Physiol.* **2011**, *157*, 1528–1536. [CrossRef]
44. Sytar, O.; Zivcak, M.; Bruckova, K.; Brestic, M.; Hemmerich, I.; Rauh, C.; Simko, I. Shift in accumulation of flavonoids and phenolic acids in lettuce attributable to changes in ultraviolet radiation and temperature. *Sci. Hortic.* **2018**, *239*, 193–204. [CrossRef]
45. Son, K.-H.; Lee, J.-H.; Oh, Y.; Kim, D.; Oh, M.-M.; In, B.-C. Growth and Bioactive Compound Synthesis in Cultivated Lettuce Subject to Light-quality Changes. *HortScience* **2017**, *52*, 584–591. [CrossRef]
46. Son, K.-H.; Park, J.-H.; Kim, D.; Oh, M.-M. Leaf Shape Index, Growth, and Phytochemicals in Two Leaf Lettuce Cultivars Grown under Monochromatic Light-emitting Diodes. *Korean J. Hortic. Sci. Technol.* **2012**, *30*, 664–672. [CrossRef]
47. Li, Q.; Kubota, C. Effects of supplemental light quality on growth and phytochemicals of baby leaf lettuce. *Environ. Exp. Bot.* **2009**, *67*, 59–64. [CrossRef]
48. Taulavuori, K.; Hyöky, V.; Oksanen, J.; Taulavuori, E.; Julkunen-Tiitto, R. Species-specific differences in synthesis of flavonoids and phenolic acids under increasing periods of enhanced blue light. *Environ. Exp. Bot.* **2016**, *121*, 145–150. [CrossRef]
49. Jiao, Y.; Lau, O.S.; Deng, X.W. Light-regulated transcriptional networks in higher plants. *Nat. Rev. Genet.* **2007**, *8*, 217–230. [CrossRef]

plants

Article

Red Light Enhances the Antioxidant Properties and Growth of *Rubus hongnoensis*

Hyeon Eui Oh, Ara Yoon and Yoo Gyeong Park *

National Institute of Biological Resources (NIBR), 1008-11, Sangnam-ro, Sangnam-myeon, Miryang 50452, Korea; hy1906@korea.kr (H.E.O.); yoonara@korea.kr (A.Y.)
* Correspondence: ygpark20@korea.kr; Tel.: +82-10-7121-7182

Abstract: The purpose of this study was to determine the effect of light quality on *R. hongnoensis* growth, physiology, and antioxidant properties. Five light conditions were employed, including white (control), red (R), blue (B), combined LED of R, green (G), and B at 7:1:2 (RGB), as well as combined LED of R, G, B, and far-red (Fr) at 7:1:2:1 (RGBFr). R light had the greatest growth-promoting effect based on plant height, leaf length, leaf width, stem diameter, and leaf area. However, leaf width and root length exhibited the greatest growth under RGB. The fresh and dry weight of shoots and roots were highest under R and RGB light. Photosynthesis was highest under RGB and lowest under B. Transpiration was highest in RGBFr. Stomatal conductance and photosynthetic water use efficiency were greatest under RGBFr. Total phenol content and radical scavenging activity were highest under R, while total flavonoid content was highest under RGB. Superoxide dismutase (SOD), catalase (CAT), and ascorbate peroxidase (APX) activities were upregulated under W, whereas guaiacol peroxidase (GPX) activity was highest under RGB. The present results suggest that, among the tested light treatments, R light was most conductive for vegetative growth and antioxidant capacity in *R. hongnoensis*.

Keywords: antioxidant activity; blue LED; photomorphogenesis; photosynthesis; red LED

Citation: Oh, H.E.; Yoon, A.; Park, Y.G. Red Light Enhances the Antioxidant Properties and Growth of *Rubus hongnoensis*. *Plants* **2021**, *10*, 2589. https://doi.org/10.3390/plants10122589

Academic Editors: Mick Fuller and Hail Rihan

Received: 28 October 2021
Accepted: 23 November 2021
Published: 26 November 2021

1. Introduction

Among Rubus species, the thornberry (*Rubus hongnoensis* Nakai) is an endemic plant belonging to the subgenus Idaeobatus and sect Rosaefolii [1]. It was first collected and identified as a new species by Nakai from a region called Hongno (near Cheonjiyeon Falls, Seogwipo), which is located on Jeju Island [2]. Rubus species have been used in traditional medicine for their many medicinal properties [3]. Components isolated from representatives of the genus have been reported to exhibit various biological activities, including antioxidant, anti-inflammatory, antibacterial, and anticancer activity [4]. As a result of comparing the physiological properties of 26 wild Rubus plants (leaf), some species showed high phenolic compound content and antioxidant activity, suggesting potential use as medicine or herbal tea [5]. Wild Rubus plants (fruits) can be used as an energy source because of their high content of essential minerals and carbohydrates [6].

Light quality can indirectly affect biomass production by influencing plant morphology, architecture, and photosynthesis [7]. Red (R) light plays an important role in photosynthesis as well as the control of shoot weight, stem diameter, and leaf area [8]. Blue (B) light not only affects plant growth, leaf expansion, and stomatal opening, but also enhances chlorophyll, flavonoid, and total phenolic content as well as antioxidant capacity [9]. The combination of R and B light is optimal for the growth and development of cucumbers [10]. Although the underlying mechanism of plant growth promotion via green (G) light is unclear [11], it has been reported that the addition of G to R and B light promotes the growth of lettuce [12] and induces an effect similar to that of shade avoidance [13]. Far-red (Fr) light has a higher leaf transmittance than R light. Therefore, it is possible to produce uniform seedlings under Fr light, as it contributes to a lower variation of seedling size [14]. However, despite the many studies above, there are no studies on the effects of various light-emitting diode (LED) light sources on the growth and antioxidant activity of *R. hongnoensis*.

In this study, we hypothesized that applying R, B, G, and Fr as monochromatic light or in combination would influence gas exchange, antioxidant activity, and the growth of *R. hongnoensis*. Therefore, the effects of four different LED spectra on growth, photosynthesis, and antioxidant activity of *R. hongnoensis* were investigated. R light considerably promoted plant growth as well as antioxidant capacity.

2. Results

Plant height was greatest under R light (11.2 cm), the leaf length was 16.7 cm, leaf width was 10.4 cm, and the leaf area was 66.2 cm^2, which altogether indicated a great improvement in leaf growth when compared to parameters under other light treatment conditions (Table 1, Figure 1). Leaf length, leaf width, and leaf area were the lowest under B light (Table 1). The number of five-leaflets was highest under RGBFr light (Table 1). Chlorophyll content was greatest under R (40.7), but no significant difference was determined between treatment groups (Table 1). Stem diameter was greatest under R and lowest under B light (Table 1).

Table 1. The growth of *Rubus hongnoensis* under different light treatments on the 8th week after transplanting.

Light Quality [z]	Plant Height (cm)	Leaf			No. of Five-Leaflets	Chlorophyll (SPAD)	Stem Diameter (mm)
		Length (cm)	Width (cm)	Area (cm^2)			
W	6.8 b [y]	13.4 a	8.4 a	44.7 ab	5.7 ab	38.0	5.3 a
R	11.2 a	16.7 a	10.4 a	66.2 a	6.6 a	40.7	6.6 a
B	3.7 c	8.8 b	5.5 b	24.1 c	5.1 b	34.3	2.8 b
RGBFr	7.9 b	15.0 a	8.5 a	51.6 b	6.8 a	39.6	5.2 a
RGB	8.2 b	15.3 a	9.0 a	55.2 ab	6.4 a	39.8	4.9 a
F-test	***	***	***	***	**	NS	**

[z] Light quality included: W, white (as the control) light-emitting diodes (LEDs); R, red LEDs; B, blue LEDs; RGB, combined LEDs of R, green (G), and B at 7:1:2; and RGBFr, combined LEDs of R, G, B, and far-red (Fr) at 7:1:2:1. [y] Mean separation within columns by Duncan's multiple range test at the 5% level. NS, **, ***: Non-significant or significant at $p \leq 0.01$ or 0.001, respectively.

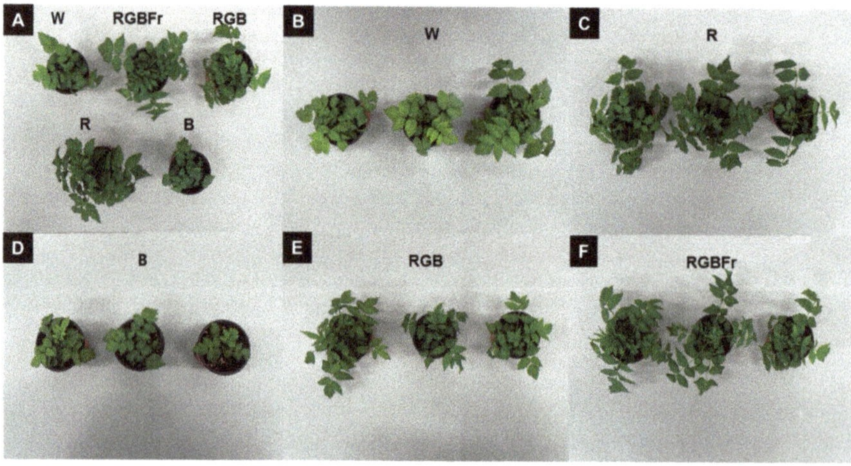

Figure 1. Comparison of shoots and roots of *Rubus hongnoensis* grown under different light-emitting diode (LED) treatment over 8 weeks: Overall view (**A**), white (**B**), red (**C**), blue (**D**), combined LEDs of R, G, and B at 7:1:2 (**E**), and combined LEDs of R, G, B, and Fr at 7:1:2:1 (**F**).

Root length (38.4 cm) was significantly higher under RGB than under the other treatment conditions (Table 2). The fresh weight (8.59 g) and dry weight (0.66 g) of roots were highest under R and RGB light, but there were no significant differences between groups (Table 2). The fresh weight and dry weight of shoots were similar between R (fresh weight-11.65 g; dry weight-2.05 g) and RGB light (fresh weight-11.9 g; dry weight-1.94 g) (Table 2).

Table 2. Root length, fresh and dry weights of *Rubus hongnoensis* under different light treatments on the 8th week after transplanting.

Light Quality [z]	Root Length (cm)	Fresh Weight (g)		Dry Weight (g)	
		Shoot	Root	Shoot	Root
W	27.1 bc [y]	5.13 a	3.68 bc	0.95 ab	0.28 bc
R	33.1 ab	11.65 a	8.34 a	2.05 a	0.64 ab
B	25.2 c	2.76 b	2.06 c	0.35 b	0.12 c
RGBFr	34.4 ab	8.29 ab	7.98 ab	1.59 a	0.52 ab
RGB	38.4 a	11.90 a	8.59 a	1.94 a	0.66 a
F-test	**	**	**	**	**

[z] See Table 1 for details on the light treatments. [y] Mean separation within columns by Duncan's multiple range test at the 5% level. **: Significant at $p \leq 0.01$.

Photosynthesis was highest under RGB (10.74 μmol CO_2 m^{-2} s^{-1}), followed by RGBFr (10.26 μmol CO_2 m^{-2} s^{-1}) and R light (10.22 μmol CO_2 m^{-2} s^{-1}), while being lowest under B light (3.44 μmol $CO_2 \cdot m^{-2}$ s^{-1}) (Figure 2A). Interestingly, photosynthesis was enhanced under exposure to R light, as observed for R, RGB, and RGBFr treatment (Figure 2A). Transpiration and stomatal conductance were highest under RGBFr and lowest under W light (Figure 2B,C). Photosynthetic water use efficiency was highest under RGB, followed by R and RGBFr (Figure 2D). Photosynthetic water use efficiency was lowest in B (Figure 2D).

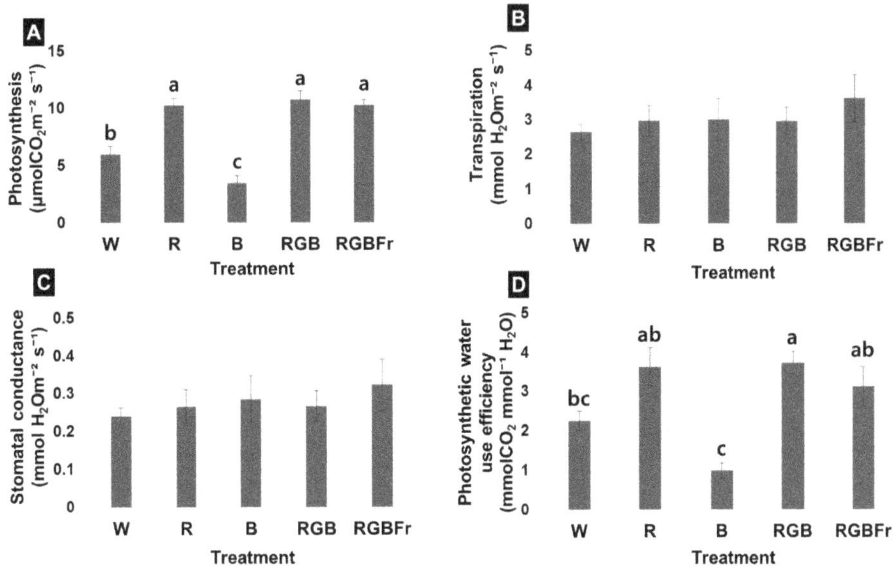

Figure 2. Effect of different light-emitting diode (LED) quality on photosynthesis (**A**), transpiration (**B**), stomatal conductance (**C**), and photosynthetic water use efficiency (**D**) in the leaves of *Rubus hongnoensis*. Light quality applied included W, white (as the control) light-emitting diodes (LEDs); R, red LEDs; B, blue LEDs; RGB, combined LEDs of R, green (G), and B at 7:1:2; and RGBFr, combined LEDs of R, G, B, and far-red (Fr) at 7:1:2:1. Data are the mean ± S.E of the 5 biological replicates. Means accompanied by different letters are significantly different ($p < 0.05$) according to the Duncan's multiple range test at 5% significance level.

Superoxide dismutase (SOD) activity of shoots was highest under B (14.9 Umg^{-1} protein), followed by RGBFr (11.3 Umg^{-1} protein) and R light (9.7 Umg^{-1} protein) (Figure 3A). SOD activity of the root was highest under W (56.2 Umg^{-1} protein) and lowest under RGB (32.1 Umg^{-1} protein) (Figure 3A). Total SOD activity was the highest under W (65.1 Umg^{-1} protein) (Figure 3A).

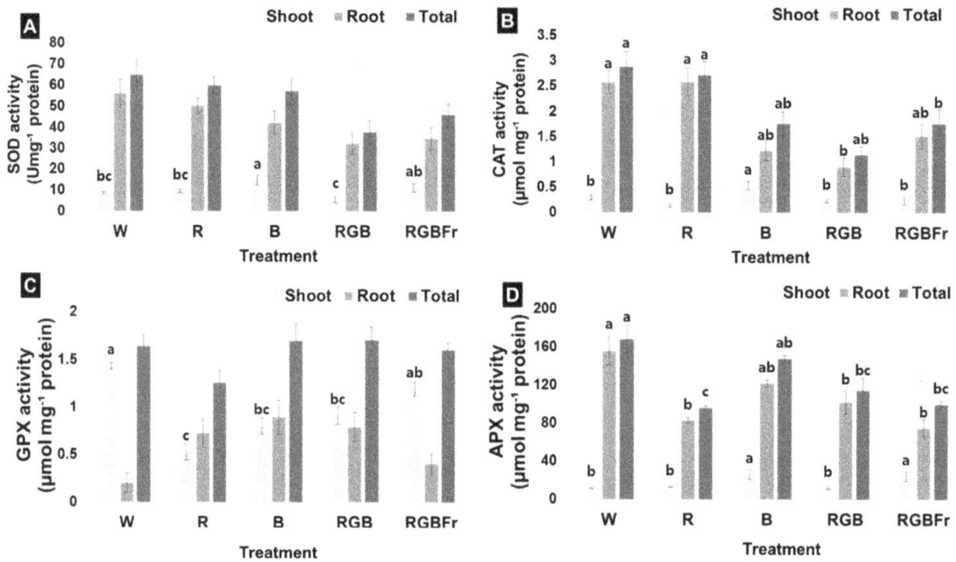

Figure 3. Superoxide dismutase (SOD) activity (**A**), catalase (CAT) activity (**B**), guaiacol peroxidase (GPX) activity (**C**), and ascorbate peroxidase (APX) activity (**D**) of *Rubus hongnoensis* as affected by light quality. Light quality applied included W, white (as the control) light-emitting diodes (LEDs); R, red LEDs; B, blue LEDs; RGB, combined LEDs of R, green (G), and B at 7:1:2; and RGBFr, combined LEDs of R, G, B, and far-red (Fr) at 7:1:2:1. Data are the mean ± S.E of the 5 biological replicates. Means accompanied by different letters are significantly different ($p < 0.05$) according to the Duncan's multiple range test at 5% significance level.

Catalase (CAT) activity in shoots was the highest under B (0.538 μmol mg^{-1} protein), while that in roots was highest under W light (2.584 μmol mg^{-1} protein) (Figure 3B). Total CAT activity was highest and similar between W (2.888 μmol mg^{-1} protein) and R (2.728 μmol mg^{-1} protein), followed by B light (1.766 μmol mg^{-1} protein) (Figure 3B).

Guaiacol peroxidase (GPX) activity of shoots was highest under W (1.435 μmol mg^{-1} protein), followed by RGBFr (1.195 μmol mg^{-1} protein), RGB (0.914 μmol mg^{-1} protein), B (0.804 μmol mg^{-1} protein), and R light (0.530 μmol mg^{-1} protein) (Figure 3C). GPX activity in the root was lowest under W light (0.207 μmol mg^{-1} protein) (Figure 3C). The total GPX activity was highest under RGB light (1.705 μmol mg^{-1} protein) (Figure 3C).

Ascorbate peroxidase (APX) activity in the shoot was similarly high under B (26.4 μmol mg^{-1} protein) and RGBFr (24.5 μmol mg^{-1} protein), followed by R (13.3 μmol mg^{-1} protein), RGB (12.2 μmol mg^{-1} protein), and W treatment (12.1 μmol mg^{-1} protein) (Figure 3D). APX activity of the root was highest under W (156.1 μmol mg^{-1} protein) as was APX activity (227.7 μmol mg^{-1} protein) (Figure 3D).

Total phenol content of the shoot was the highest under W (3.03 mg g^{-1}), followed by R, RGB, RGBFr, and lowest under B light (1.26 mg g^{-1}) (Figure 4A). In the root, phenol content was highest under R (1.547 mg g^{-1}) and lowest under B light (0.75 mg g^{-1}) (Figure 4A). Total phenol content was highest under R (4.44 mg g^{-1}), followed by W (4.01 mg g^{-1}), RGB (2.98 mg g^{-1}), and RGBFr (2.36 mg g^{-1}), again being lowest under B light (2.05 mg g^{-1}) (Figure 4A).

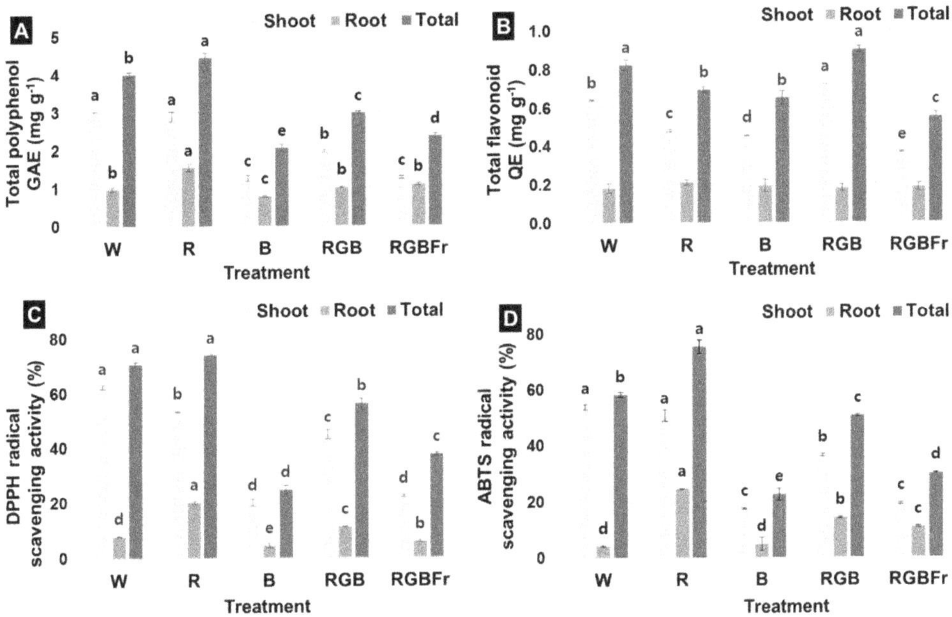

Figure 4. Total phenol content (**A**), total flavonoid content (**B**), 2,2-Diphenyl-1-picrylhydrazyl (DPPH) radical scavenging (**C**), and 2,2′-azinobis-(3-ethyl-benzothiazoline)-sulfonic acid (ABTS) radical scavenging activity (**D**) in *R. hongnoensis* was affected by light quality. Light quality applied included W, white (as the control) light-emitting diodes (LEDs); R, red LEDs; B, blue LEDs; RGB, combined LEDs of R, green (G), and B at 7:1:2; and RGBFr, combined LEDs of R, G, B, and far-red (Fr) at 7:1:2:1. Data are the mean ± S.E of the 5 biological replicates. Means accompanied by different letters are significantly different ($p < 0.05$) according to the Duncan's multiple range test at 5% significance level.

Total flavonoid content of the shoot was the highest under RGB (0.72 mg g^{-1}), followed by W (0.64 mg g^{-1}), R (0.48 mg g^{-1}), B (0.46 mg g^{-1}), and RGBFr light (0.37 mg g^{-1}) (Figure 4B). Root flavonoid content was highest under R (0.21 mg g^{-1}), but there was no significant difference between treatment groups (Figure 4B). Total flavonoid content was higher under RGB (0.9 mg g^{-1}) than W (0.82 mg g^{-1}) and lowest under RGBFr light (0.55 mg g^{-1}) (Figure 4B).

The 2,2-Diphenyl-1-picrylhydrazyl (DPPH) radical scavenging activity of shoots was highest under W (62.6%), followed by R (53.6%), RGB (44.8%), RGBFr (22.5%), and lowest under B light (20.2%) (Figure 4C). In contrast, DPPH activity in roots was highest under R (20.3%) and lowest under B (4.6%) (Figure 4C). Total DPPH activity was highest under R (73.9%), followed by W (70.6%), RGB (56.3%), RGBFr (37.6%), and lowest under B light (24.8%) (Figure 4C).

The 2,2′-azinobis-(3-ethyl-benzothiazoline)-sulfonic acid (ABTS) radical scavenging activity of shoots was highest under W (54.2%), followed by R (50.7%), RGB (36.4%), RGBFr (19.2%), and lowest under B light (17.5%) (Figure 4D). Root ABTS activity was highest under R (24.6%) and lowest under W (4.2%) (Figure 4D). Total ABTS activity was highest under R (75.3%), followed by W (58.4%), RGB (50.6%), RGBFr (29.9%), and lowest under B light (22.5%) (Figure 4D).

3. Discussion

3.1. Morphogenesis

In the present study, leaf growth parameters, such as length, width, and leaf area, were highest under R as compared to the control treatment. Ohashi-Kaneko et al. [15] found that *Brassica campestris* leaf growth was also greatest under R light. According to Wu and

Lin [16], *Protea cynaroides* plantlets grown under R LEDs produced a significantly higher number of new leaves compared to those grown under other LED treatments. The fresh and dry weights of *R. hongnoensis* shoots were similar under R and RGB light conditions in the present study. Heo et al. [17] reported that the fresh and dry weights of grape increased under R light. Lee et al. [18] demonstrated the shoot and root growth-promoting effect under different light conditions, including R light, which was also observed in the present experiments.

Pecháčková [19] noted that root growth and development can be altered by light quality. Root growth was enhanced under R, RGB, and RGBFr treatment in the current study, indicating that light sources containing R had a favorable effect on this parameter. Similar observations were reported in *Gossypium hirutum* [20], and *Chrysanthemum morifolium* [21], where R LEDs were found to stimulate root formation. According to Wu and Lin [16], *Protea cynaroides* root growth was highest under R light. Simlat et al. [21] also reported that R light had a positive effect on root growth.

In the present study, all evaluated growth parameters of *R. hongnoensis* were lowest under B, suggesting that B light did not promote growth. A similar result was reported by Heo et al. [17] who demonstrated the growth inhibitory effect of B light in the sprouts of some greenhouse crops. In addition, Wu et al. [22] reported that the elongation of *Solanum lycopersicum* was inhibited by B light.

3.2. Photosynthesis

Light provides energy for photosynthesis, and thus light quality has major influence on the process [9]. In the present study, photosynthesis was greatest under RGB compared the control W light condition, followed by RGBFr and R, while B light treatment resulted in the lowest photosynthetic activity. Similar results were reported by Kim et al. [12], who demonstrated that RGB treatment enhanced lettuce growth. Although the combination of R and B LEDs has great potential use as a light source for enhancing photosynthesis, plants have adapted to utilize a wider spectrum of light to control photomorphogenesis [23]. Emerson and Rabinowitch [24] reported that photosynthesis was enhanced under the concurrent application of two light beams of different quality.

The transpiration rate and stomatal conductance increased under RGBFr, RGB, and B light in the present study. A similar result was reported by Yorio et al. [25], who reported that stomatal opening was stimulated in the leaves of lettuce grown under R LED light supplemented with B light. B light strongly affects plant growth and development, including leaf size, stomatal opening, and photosynthesis [26]. In the present study, the photosynthetic efficiency of *R. hongnoensis* leaves was improved under all light sources containing R light, while monochromatic B light had a negative effect on photosynthesis. Similar results were reported that R light was important for photosynthetic apparatus development as it enhanced starch accumulation in various plant species by inhibiting the translocation of photosynthates out of leaves [27,28].

3.3. Antioxidant

Light is known to affect not only plant growth and development but also the biosynthesis of primary and secondary metabolites [29]. Various LED radiation treatments have been reported to promote antioxidant enzyme activity [21]. In the present study, we also observed that different LED light treatment affected the activity of reactive oxygen species-scavenging enzymes in *R. hongnoensis*. SOD, commonly called metalloenzyme, decomposes two highly reactive O_2^- to produce H_2O_2 and O_2 [30]. CAT, APX, and GPX reduce H_2O_2 to water and molecular oxygen [30]. In *R. hongnoensis* shoots, the production of H_2O_2 and O_2 increased as the SOD activity increased under B light, and the CAT and APX activities were also increased to degrade the increased H_2O_2. B LED treatment was promoted on ROS-scavenging enzyme activity (SOD, CAT, and APX), opposite to that of R treatment [31]. It is interesting that the GPX activity of root under W and RGBFr is lower than shoot. Although the lower dry weight of the roots under the W treatment might suggest that the root growth in the container was less stressful, it is not clear why the GPX activity was lower in the RGBFr treatment.

In addition to antioxidant enzymes, plants produce various antioxidant compounds, such as phenols, in an attempt to respond and adapt to various biotic as well as abiotic stressors that would otherwise damage the photosynthetic apparatus [32]. Phenols are abundant secondary metabolites in plants and act as natural antioxidants with a wide range of biological activities, including antioxidant, anticancer, antibacterial, and anti-inflammatory activities [33]. In the present study, B light clearly suppressed the accumulation of phenol contents in *R. hongnoensis*. Zheng and Van Labeke [34] reported that total phenol content in the leaves of *Dendranthema morifolium* was suppressed under B light, although this effect depended on the cultivar. B, R, and FR wavelengths also regulate the biosynthesis of phenol contents in a direct or indirect manner through signaling, which leads to the expression of key enzymes, or through upregulating the production of shikimic acid, which is a precursor of phenol contents [35]. Moreover, the activity of phenylalanine ammonia lyase (PAL), a major enzyme involved in phenolic biosynthesis, is known to be regulated by light quality [36]. The total phenol content of shoots and roots in the present study was highest under R, being higher in shoots. R LEDs have been widely employed as an alternative source of illumination for in vitro survival as well as for enhancing metabolite production in medicinally important plants [37]. Shohael et al. [38] demonstrated that R light induced the synthesis of secondary metabolites in *Eleutherococcus senticosus*, resulting in the highest total phenol content, as opposed to B light under which phenol content was lowest.

Plant growth and flavonoid biosynthesis are stimulated by multiple factors, including the specific characteristics of visible light quality [39]. Our study confirmed that RGB light resulted in the greatest accumulation of flavonoid contents in *R. hongnoensis* leaves, while the lowest accumulation was under B light. In contrast, Taulavuori et al. [39] found that G and B light increased the flavonoid content in tobacco leaves and red leaf lettuce. B LED exposure was also reported to increase the total flavonoid content of *C. paliurus* leaves by 37.7% and their quercetin content by 184.6% when compared to W LED after 60 days of treatment [40]. According to a study by Ouzounis et al. [41], the flavonoid content in tomatoes increased under 12% R supplementation, depending on the genotype. The high flavonoid content generated under RGB light conditions in the present study is believed to be due to the synergetic effect of R, G, and B light.

DPPH and ABTS were employed for the assessment of phenol content antioxidant capacity in the present work [42]. DPPH is a stable free radical that becomes a stable diamagnetic molecule by accepting electrons or hydrogen radicals [42]. In this study, the antioxidant activity of *R. hongnoensis* differed depending on light quality. Similarly, Shiga et al. [43] reported that the DPPH free radical scavenging activity of *Ocimum basilicum* L. was greater under R light than under B light. The antioxidant capacities determined via both methods (DPPH and ABTS) were in an agreement with the total phenol content of plants determined in our study. The present results are consistent with those of several other studies [44]. Chen et al. [45] reported a significant positive correlation between the total phenol content of persimmon and DPPH as well as ABTS radical scavenging activity. Therefore, it was concluded that the phenol contents of *R. hongnoensis* were increased under R light, thus improving antioxidant capacity.

4. Materials and Methods

4.1. Plant Materials and Growth Conditions

Germinating *R. hongnoensis* seeds (NIBRGR0000624110) provided by the National Institute of Biological Resources were used. Among the germinated seedlings, individuals with two or more true leaves were selected and transplanted into 72-cell plug trays with a commercial medium (Baroker; Seoul Bio Co., Ltd., Eumseong, Korea) and were grown under LED light for 8 weeks. After 4 weeks of LED light treatment, plants were transplanted into pots (10 cm in diameter), and after another 6 weeks, plants with a height of 8 cm or more were transplanted into pots (20 cm in diameter). During the experiment, air humidity in the cultivation room was maintained at 60%, the temperature was 23 °C, and the photoperiod was 12/12 h (dark/light), with tap water irrigation twice a week.

4.2. Light Treatment

The light source used in the experiment was an LED lamp (1280W, KNP LED, Daegu, Korea). Five types of single and mixed light sources were applied in different experimental treatment groups, including W light as a control, in addition to R (660 nm), B (444 nm), G (519 nm), and Fr (732 nm) LEDs. The light intensity was maintained at 200 μmol m^{-2} s^{-1} PPFD, and all treatments included 404 nm 20 μmol m^{-2} s^{-1} PPFD. The average PPFD was measured with a light meter (MK350, UPRTek, Jhunan, Taiwan) at a distance of 32 cm above the bench top. The spectral distribution of light in the experiment was measured using a spectroradiometer (MQ-200, Apogee, Logan, USA) 32 cm above the bench top, at 1-nm wavelength intervals. The spectral distribution and characteristics measured at three locations within the plant growing area for each treatment are shown in Figure 5. A completely randomized block design with 3 replications and 5 seedlings per treatment was used in the experiment. The treatment of locations in a controlled environment were randomly mixed between replications in order to minimize positional effects.

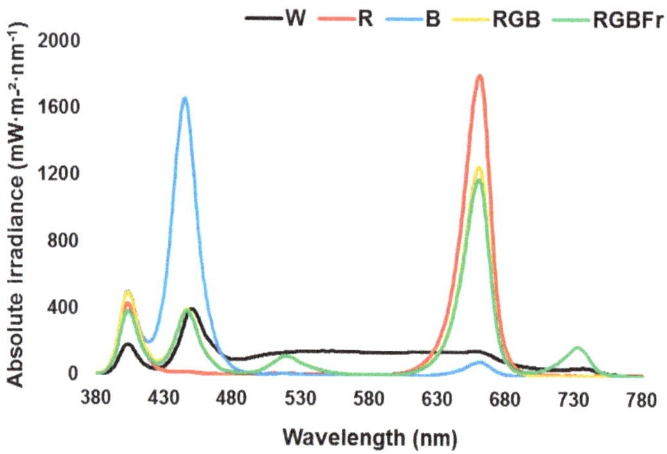

Figure 5. The spectral distribution of lights used in the closed walk-in growth chamber. Light quality used included W, white (as the control) light-emitting diodes (LEDs); R, red LEDs; B, blue LEDs; RGB, combined LEDs of R, green (G), and B at 7:1:2; and RGBFr, combined LEDs of R, G, B, and far-red (Fr) at 7:1:2:1.

4.3. Growth Characteristics

Four weeks after treatment, plant height, leaf length, leaf width, leaf area, number of five-leaflets, root length, as well as the fresh and dry weights of shoots and roots were measured. Leaf area was measured using a leaf area meter (LI-3100, LICOR Inc., Lincoln, NE, USA), and the leaf area per compound leaf was determined. The number of five-leaflets was measured by counting the number of leaflets per compound leaf. Root length was measured as the length of the longest root. Fresh weight was measured using an electronic scale (CATY224, CAS Co., Seoul, Korea). Dry weight was measured after drying the tissues for 48 h at 70 °C in a drying oven (SJ-202DM, Sejong Scientific Co., Ltd., Bucheon, Korea).

4.4. Photosynthesis Measurements

Photosynthesis was measured in the terminal leaflet of the largest compound leaf using a portable photosynthesis system (Portable Photosynthesis system, Li-6800, LICOR Inc., Lincoln, NE, USA). The net photosynthesis, stomatal conductance, transpiration, and photosynthetic water use efficiency were calculated. Photosynthesis measurements were conducted 4 weeks after light treatment and immediately before final growth irradiation. Measurements were performed in survey mode and were repeated four times per treatment condition. The measurement conditions were 600 $\mu mol \cdot s^{-1}$ inflow air flow into the chamber, 25 °C temperature, 70% relative humidity, 3 cm^2 leaf area, and 400 $\mu mol\ mol^{-1}$ CO_2. The light source in the chamber was removed to determine photosynthetic capacity under light conditions given to the experimental treatment.

4.5. Enzymatic Antioxidants

For the measurement of antioxidant activity, 100 mg fresh weight was added to 1 mL of 50 mM phosphate buffer with a pH of 7.0 containing 0.1 mM EDTA. The sample was extracted by bolting for 10 s and was then centrifuged at 4 °C and 13000 rpm for 20 min. One hundred milligrams fresh weight of *R. hongnoensis* were added to 1 mL of 50 mM phosphate buffer (pH 7.0) containing 0.1 mM EDTA, vortexed for 10 s, and centrifuged at 4 °C and 13,000 rpm for 20 min. SOD activity was measured via the method described by Alici and Arabaci [46]. SOD riboflavin was prepared by adding 2 g pvp, 50 µL triton-X, and 0.314 g riboflavin to 100 mL of 50 mM phosphate buffer. The SOD reaction mixture was prepared by adding 2 mM EDTA, 9.9 mM L-methionine, and 55 µM NBT to 100 mL of distilled water. In 1.25 mL of SOD reaction mixture, 50 µL of enzyme extract, and 200 µL of SOD riboflavin were mixed and reacted for 15 min under light at room temperature, and absorbance was measured at 560 nm. As a control, a reactant that was not irradiated with light was used.

CAT activity was measured with some modifications to the method described by Aebi [47]. The reaction mixture contained 100 µL of enzyme extract, 150 µL of H_2O_2, and 1.25 mL of 50 mM phosphate buffer. The change in absorbance at 240 nm was determined at 30-s intervals for 3 min. The molar extinction coefficient of H_2O_2 was [40 $mmol^{-1}$ cml^{-1}] at 240 nm.

GPX activity was measured as per the method described by Sadasivam and Manickam, with some modifications [48]. The reaction mixture contained 100 µL enzyme extract, 20 mM guaiacol 100 µL, 30 mM H_2O_2 50 µL, and 1.25 mL 50 mM phosphate buffer. The change in absorbance at 436 nm was them measured at 30-s intervals for 3 min. The molar extinction coefficient of H_2O_2 was 25 $mmol^{-1}$ cml^{-1}] at 436 nm.

APX activity was measured as per the method described by Chen and Asada with some modifications [49]. The reaction mixture contained 100 µL of enzyme extract, 50 µL of 100 mM H_2O_2, and 1.3 mL of 50 mM phosphate buffer containing 0.6 mM ascorbic acid. The change in absorbance at 290 nm was measured for 33 min. The molar extinction coefficient of H_2O_2 was [2.8 $mmol^{-1}$ cml^{-1}] at 290 nm.

4.6. Extract Preparation

For the antioxidant assay, leaves, stems, and roots were harvested after 8 weeks of light treatment. The samples obtained by dividing the leaves, stems (shoots), and roots (roots) were frozen in liquid nitrogen and stored in a cryogenic freezer (UniFreezerU80, Daehan Scientific Co. Ltd., Wonju, Korea) at −80 °C. Frozen samples were ground in a mortar and used for the analysis. To prepare the sample extract, 100 mg of the sample and 1 mL of 50% methanol were mixed and stored for 6 h, followed by centrifugation (5424R, Eppendorf, Hamburg, Germany) at 13,000 rpm and 4 °C for 20 min.

4.7. Total Phenol Content and Flavonoid Content

The total phenol content of extracts was measured as per the Folin–Ciocalteu method [50]. To prepare the sample, 100 mg of the sample and 1 mL of 50% methanol were mixed and stored for 6 h, followed by centrifugation (5424R, Eppendorf, Hamburg, Germany) at 13,000 rpm and 4 °C for 20 min. Next, 500 µL of 2% Na_2CO_3 was added to a mixture of 450 µL distilled water, 250 µL 50% Folin–Ciocalteu solution, and 100 µL sample extract diluted 10 times, followed by incubation in the dark for 30 min. The absorbance was measured at 765 nm using a UV spectrophotometer (UV–1280, Shimadzu, Japan). The total phenol content was calculated using gallic acid as the standard.

4.8. Total Flavonoid Content

The total flavonoid content of the extract was measured according to the method of Kumaran and Karunakaran [51]. Fifty microliters of the extract was added to a mixture of 450 μL 80% methanol and 500 μL 2% $AlCl_3$, vortexed for 2 s, and then reacted at room temperature for 30 min. After the reaction was completed, the absorbance at 415 nm was measured using a spectrophotometer. The total flavonoid content was calculated using quercetin as the standard.

4.9. DPPH Radical Scavenging Assay

The DPPH radical scavenging ability of the extract was determined according to the method of Blois [52]. DPPH was measured by adding 0.2 mM DPPH 800 μL to 200 μL of extracted sample, allowing it to react in the dark for 30 min, and then measuring the absorbance at 520 nm. Methanol was added instead of the sample extract as a control. Ascorbic acid was used as the positive control instead of the sample extract.

4.10. ABTS Radical Scavenging Assay

The ABTS radical scavenging ability of the extract was determined according to the method of Re et al. [53]. To prepare ABTS reagent, 7.4 mM ABTS and 2.6 mM potassium persulfate were mixed and stored in the dark for 24 h, and then the absorbance at 735 nm was adjusted to 0.7 ± 0.05. After adding 10 μL of the sample extract to 1 mL of ABTS reagent, the mixture was allowed to react for 10 min. Absorbance was measured at 735 nm. Methanol was added as a control instead of the sample. The scavenging activity (%) for cationic radicals and free radicals was calculated as [1−(sample/control)] × 100, and compared with L-ascorbic acid, a control material.

4.11. Experimental Design and Statistical Analysis

Data were analyzed for statistical significance using the SAS (Statistical Analysis System, V. 9.4, Cary, NC, USA) program. The experimental results were subjected to analysis of variance (ANOVA) and Duncan's multiple range test.

5. Conclusions

R light treatment resulted in greater growth and promotion of antioxidant activity in *R. hongnoensis*. The results of this study demonstrated that R light increases the total phenol content as well as radical scavenging capacity. Further studies exploring the optimal light intensity and irradiation time are still needed in order to improve the application of R light technology for the promotion of *R. hongnoensis* antioxidant capacity.

Author Contributions: Conceptualization, Y.G.P. and H.E.O.; methodology, Y.G.P. and H.E.O.; formal analysis, H.E.O. and A.Y.; data curation, Y.G.P.; writing—original draft preparation, H.E.O., A.Y. and Y.G.P.; writing—review and editing, Y.G.P. All authors have read and agreed to the published version of the manuscript.

Funding: This work was supported by a grant from the National Institute of Biological Resources (NIBR), funded by the Ministry of Environment (MOE) of the Republic of Korea (NIBR202115101).

Institutional Review Board Statement: Not applicable.

Informed Consent Statement: Not applicable.

Data Availability Statement: The authors will make the results available if requested.

Conflicts of Interest: The authors declare no conflict of interest.

References

1. Ma, J.; Park, C.-W. The Genera of Vascular Plants of Korea. *TAXON* **2008**, *57*, 681. [CrossRef]
2. Nakai, T. Plantae novae Coreanae et Japonicae II. *Repert. Spec. Nov. Regni Veg.* **1914**, *13*, 277. [CrossRef]
3. Patel, A.V.; Rojas-Vera, J.; Dacke, C.G. Therapeutic constituents and actions of Rubus species. *Curr. Med. Chem.* **2004**, *11*, 1501–1512. [CrossRef] [PubMed]
4. Lechowicz, K.; Wrońska-Pilarek, D.; Bocianowski, J.; Maliński, T. Pollen morphology of Polish species from the genus Rubus L. (Rosaceae) and its systematic importance. *PLoS ONE* **2020**, *15*, e0221607. [CrossRef]
5. Oszmiański, J.; Wojdyło, A.; Nowicka, P.; Teleszko, M.; Cebulak, T.; Wolanin, M. Determination of Phenolic Compounds and Antioxidant Activity in Leaves from Wild Rubus L. Species. *Molecules* **2015**, *20*, 4951–4966. [CrossRef]
6. Ahmad, M.; Masood, S.; Sultana, S.; Ben Hadda, T.; Bader, A.; Zafar, M. Report: Antioxidant and nutraceutical value of wild medicinal Rubus berries. *Pak. J. Pharm. Sci.* **2015**, *28*, 241–247.
7. Van Ieperen, W. Plant Morphological and Developmental Responses to Light Quality in a Horticultural Context. *Acta Hortic.* **2012**, 131–139. [CrossRef]
8. Xu, Y.; Liang, Y.; Yang, M. Effects of Composite LED Light on Root Growth and Antioxidant Capacity of Cunninghamia lanceolata Tissue Culture Seedlings. *Sci. Rep.* **2019**, *9*, 1–9. [CrossRef]
9. Wang, H.; Gu, M.; Cui, J.; Shi, K.; Zhou, Y.; Yu, J. Effects of light quality on CO_2 assimilation, chlorophyll-fluorescence quenching, expression of Calvin cycle genes and carbohydrate accumulation in Cucumis sativus. *J. Photochem. Photobiol. B Biol.* **2009**, *96*, 30–37. [CrossRef] [PubMed]
10. Li, H.; Cheng, Z. Hoagland nutrient solution promotes the growth of cucumber seedlings under light-emitting diode light. *Acta Agric. Scand. Sect. B Plant. Soil Sci.* **2014**, *65*, 74–82. [CrossRef]
11. Jishi, T. LED Lighting Technique to Control Plant Growth and Morphology. In *Smart Plant Factory*; Springer: Singapore, 2018; pp. 211–222. [CrossRef]
12. Kim, H.-H.; Goins, G.D.; Wheeler, R.M.; Sager, J.C. Stomatal Conductance of Lettuce Grown under or Exposed to Different Light Qualities. *Ann. Bot.* **2004**, *94*, 691–697. [CrossRef] [PubMed]
13. Zhang, T.; Maruhnich, S.A.; Folta, K.M. Green Light Induces Shade Avoidance Symptoms. *Plant. Physiol.* **2011**, *157*, 1528–1536. [CrossRef] [PubMed]
14. Shibuya, T.; Endo, R.; Kitaya, Y.; Hayashi, S. Growth Analysis and Photosynthesis Measurements of Cucumber Seedlings Grown under Light with Different Red to Far-red Ratios. *HortScience* **2016**, *51*, 843–846. [CrossRef]
15. Ohashi-Kaneko, K.; Takase, M.; Kon, N.; Fujiwara, K.; Kurata, K. Effect of Light Quality on Growth and Vegetable Quality in Leaf Lettuce, Spinach and Komatsuna. *Environ. Control. Biol.* **2007**, *45*, 189–198. [CrossRef]
16. Wu, H.-C.; Lin, C.-C. Red Light-emitting Diode Light Irradiation Improves Root and Leaf Formation in Difficult-to-propagate *Protea cynaroides* L. Plantlets In Vitro. *HortScience* **2012**, *47*, 1490–1494. [CrossRef]
17. Heo, J.W.; Shin, K.S.; Kim, S.K.; Paek, K.Y. Light quality affectsin Vitro growth of grape 'Teleki 5BB'. *J. Plant. Biol.* **2006**, *49*, 276–280. [CrossRef]
18. Lee, J.G.; Oh, S.S.; Cha, S.H.; Jang, Y.A.; Kim, S.Y.; Um, Y.C.; Cheong, S.R. Effects of red/blue light ratio and short-term light quality conversion on growth and anthocyanin contents of baby leaf lettuce. *J. Bio-Environ.* **2010**, *19*, 351–359.
19. Pecháčková, S. Root response to above-ground light quality–Differences between rhizomatous and non-rhizomatous clones of *Festuca rubra*. *Plant Ecol.* **1999**, *141*, 67–77. [CrossRef]
20. Li, H.; Xu, Z.; Tang, C. Effect of light-emitting diodes on growth and morphogenesis of upland cotton (*Gossypium hirsutum* L.) plantlets in vitro. *Plant Cell Tissue Organ Cult.* **2010**, *103*, 155–163. [CrossRef]
21. Simlat, M.; Ślęzak, P.; Moś, M.; Warchoł, M.; Skrzypek, E.; Ptak, A. The effect of light quality on seed germination, seedling growth and selected biochemical properties of Stevia rebaudiana Bertoni. *Sci. Hortic.* **2016**, *211*, 295–304. [CrossRef]
22. Wu, Q.; Su, N.; Shen, W.; Cui, J. Analyzing photosynthetic activity and growth of Solanum lycopersicum seedlings exposed to different light qualities. *Acta Physiol. Plant.* **2014**, *36*, 1411–1420. [CrossRef]
23. Park, Y.G.; Park, J.E.; Hwang, S.J.; Jeong, B.R. Light source and CO_2 concentration affect growth and anthocyanin content of lettuce under controlled environment. *Hortic. Environ. Biotechnol.* **2012**, *53*, 460–466. [CrossRef]
24. Emerson, R.; Rabinowitch, E. Red Drop and Role of Auxiliary Pigments in Photosynthesis. *Plant Physiol.* **1960**, *35*, 477–485. [CrossRef] [PubMed]
25. YYorio, N.C.; Goins, G.D.; Kagie, H.R.; Wheeler, R.M.; Sager, J.C. Improving Spinach, Radish, and Lettuce Growth under Red Light-emitting Diodes (LEDs) with Blue Light Supplementation. *HortScience* **2001**, *36*, 380–383. [CrossRef]
26. Gajc-Wolska, J.; Kowalczyk, K.; Przybysz, A.; Mirgos, M.; Orliński, P. Photosynthetic Efficiency and Yield of Cucumber (*Cucumis sativus* L.) Grown under HPS and LED Lighting in Autumn–Winter Cultivation. *Plants* **2021**, *10*, 2042. [CrossRef]
27. Yang, X.; Xu, H.; Shao, L.; Li, T.; Wang, Y.; Wang, R. Response of photosynthetic capacity of tomato leaves to different LED light wavelength. *Environ. Exp. Bot.* **2018**, *150*, 161–171. [CrossRef]
28. Kobayashi, K.; Amore, T.; Lazaro, M. Light-Emitting Diodes (LEDs) for Miniature Hydroponic Lettuce. *Opt. Photon. J.* **2013**, *3*, 74–77. [CrossRef]
29. Hwang, C.H.; Park, Y.G.; Jeong, B.R. Changes in content of total polyphenol and activities of antioxidizing enzymes in Perilla frutescens var. acuta Kudo and Salvia plebeia R. Br. as affected by light intensity. *Hortic. Environ. Biotechnol.* **2014**, *55*, 489–497. [CrossRef]

30. Dewir, Y.; Chakrabarty, D.; Ali, M.; Hahn, E.; Paek, K.Y. Lipid peroxidation and antioxidant enzyme activities of Euphorbia millii hyperhydric shoots. *Environ. Exp. Bot.* **2006**, *58*, 93–99. [CrossRef]
31. Yu, W.; Liu, Y.; Song, L.; Jacobs, D.; Du, X.; Ying, Y.; Shao, Q.; Wu, J. Effect of Differential Light Quality on Morphology, Photosynthesis, and Antioxidant Enzyme Activity in Camptotheca acuminata Seedlings. *J. Plant. Growth Regul.* **2016**, *36*, 148–160. [CrossRef]
32. Bantis, F. Light Spectrum Differentially Affects the Yield and Phytochemical Content of Microgreen Vegetables in a Plant Factory. *Plants* **2021**, *10*, 2182. [CrossRef] [PubMed]
33. Ye, F.; Liang, Q.; Li, H.; Zhao, G. Solvent effects on phenolic content, composition, and antioxidant activity of extracts from florets of sunflower (*Helianthus annuus* L.). *Ind. Crop. Prod.* **2015**, *76*, 574–581. [CrossRef]
34. Zheng, L.; Van Labeke, M.-C. Chrysanthemum morphology, photosynthetic efficiency and antioxidant capacity are differentially modified by light quality. *J. Plant Physiol.* **2017**, *213*, 66–74. [CrossRef] [PubMed]
35. Giménez, A.; Martínez-Ballesta, M.; Egea-Gilabert, C.; Gómez, P.; Artés-Hernández, F.; Pennisi, G.; Orsini, F.; Crepaldi, A.; Fernández, J. Combined Effect of Salinity and LED Lights on the Yield and Quality of Purslane (*Portulaca oleracea* L.) Microgreens. *Horticulturae* **2021**, *7*, 180. [CrossRef]
36. Hashim, M.; Ahmad, B.; Drouet, S.; Hano, C.; Abbasi, B.; Anjum, S. Comparative Effects of Different Light Sources on the Production of Key Secondary Metabolites in Plants In Vitro Cultures. *Plants* **2021**, *10*, 1521. [CrossRef]
37. Hogewoning, S.; Trouwborst, G.; Maljaars, H.; Poorter, H.; Van Ieperen, W.; Harbinson, J. Blue light dose-responses of leaf photosynthesis, morphology, and chemical composition of Cucumis sativus grown under different combinations of red and blue light. *J. Exp. Bot.* **2010**, *61*, 3107–3117. [CrossRef] [PubMed]
38. Shohael, A.; Ali, M.; Yu, K.; Hahn, E.; Islam, R.; Paek, K. Effect of light on oxidative stress, secondary metabolites and induction of antioxidant enzymes in Eleutherococcus senticosus somatic embryos in bioreactor. *Process. Biochem.* **2006**, *41*, 1179–1185. [CrossRef]
39. Taulavuori, K.; Hyöky, V.; Oksanen, J.; Taulavuori, E.; Julkunen-Tiitto, R. Species-specific differences in synthesis of flavonoids and phenolic acids under increasing periods of enhanced blue light. *Environ. Exp. Bot.* **2016**, *121*, 145–150. [CrossRef]
40. Liu, Y.; Fang, S.; Yang, W.; Shang, X.; Fu, X. Light quality affects flavonoid production and related gene expression in Cyclocarya paliurus. *J. Photochem. Photobiol. B Biol.* **2018**, *179*, 66–73. [CrossRef]
41. Ouzounis, T.; Heuvelink, E.; Ji, Y.; Schouten, H.; Visser, R.; Marcelis, L. Blue and red LED lighting effects on plant biomass, stomatal conductance, and metabolite content in nine tomato genotypes. *Acta Hortic.* **2016**, 251–258. [CrossRef]
42. Alma, M.H.; Mavi, A.; Yildirim, A.; Digrak, M.; Hirata, T. Screening Chemical Composition and in Vitro Antioxidant and Antimicrobial Activities of the Essential Oils from *Origanum syriacum* L. Growing in Turkey. *Biol. Pharm. Bull.* **2003**, *26*, 1725–1729. [CrossRef]
43. Shiga, T.; Shoji, K.; Shimada, H.; Hashida, S.-N.; Goto, F.; Yoshihara, T. Effect of light quality on rosmarinic acid content and antioxidant activity of sweet basil, *Ocimum basilicum* L. *Plant Biotechnol.* **2009**, *26*, 255–259. [CrossRef]
44. Denoya, G.I.; Pataro, G.; Ferrari, G. Effects of postharvest pulsed light treatments on the quality and antioxidant properties of persimmons during storage. *Postharvest Biol. Technol.* **2020**, 160. [CrossRef]
45. Chen, X.N.; Fan, J.F.; Yue, X.; Wu, X.R.; Li, L.T. Radical Scavenging Activity and Phenolic Compounds in Persimmon (*Diospyros kaki* L. cv. Mopan). *J. Food Sci.* **2007**, *73*, C24–C28. [CrossRef] [PubMed]
46. Alici, E.H.; Arabaci, G. Determination of SOD, POD, PPO and CAT Enzyme Activities in *Rumex obtusifolius* L. *Annu. Res. Rev. Biol.* **2016**, *11*, 1–7. [CrossRef]
47. Aebi, H. Catalase. In *Methods of Enzymatic Analysis*; Academic Press: New York, NY, USA, 1974; pp. 673–684.
48. Sadasivam, S.; Manickam, A. *Biochemical Methods*; New Age International (P) Limited: New Delhi, India, 1996; Volume 2, pp. 124–126.
49. Chen, G.-X.; Asada, K. Ascorbate Peroxidase in Tea Leaves: Occurrence of Two Isozymes and the Differences in Their Enzymatic and Molecular Properties. *Plant. Cell Physiol.* **1989**, *30*, 987–998. [CrossRef]
50. Zhou, Y.; Singh, B.R. Red light stimulates flowering and anthocyanin biosynthesis in American cranberry. *Plant Growth Regul.* **2002**, *38*, 165–171. [CrossRef]
51. Kumaran, A.; Karunakaran, R.J. In vitro antioxidant activities of methanol extracts of five Phyllanthus species from India. *LWT Food Sci. Technol.* **2007**, *40*, 344–352. [CrossRef]
52. Blois, M.S. Antioxidant Determinations by the Use of a Stable Free Radical. *Nature* **1958**, *181*, 1199–1200. [CrossRef]
53. Re, R.; Pellegrini, N.; Proteggente, A.; Pannala, A.; Yang, M.; Rice-Evans, C. Antioxidant activity applying an improved ABTS radical cation decolorization assay. *Free Radic. Biol. Med.* **1999**, *26*, 1231–1237. [CrossRef]

plants

MDPI

Article

The Impact of LED Lighting Spectra in a Plant Factory on the Growth, Physiological Traits and Essential Oil Content of Lemon Balm (*Melissa officinalis*)

Hail Z. Rihan [1,2,*], Naofel Aljafer [1], Marwa Jbara [3], Lynn McCallum [3], Sabine Lengger [4] and Michael P. Fuller [1]

1 School of Biological and Marine Sciences, Faculty of Science and Engineering, University of Plymouth, Plymouth PL4 8AA, UK; naofel.aljafer@plymouth.ac.uk (N.A.); mfuller@plymouth.ac.uk (M.P.F.)
2 Phytome Life Sciences, Launceston PL15 7AB, UK
3 School of Biomedical Sciences, Faculty of Health, University of Plymouth, Plymouth PL4 8AA, UK; marwa.jbara@plymouth.ac.uk (M.J.); lynn.mccallum@plymouth.ac.uk (L.M.)
4 School of Geography, Earth and Environmental Sciences, Faculty of Science and Engineering, University of Plymouth, Plymouth PL4 8AA, UK; sabine.lengger@plymouth.ac.uk
* Correspondence: hail.rihan@plymouth.ac.uk; Tel.: +44-(75)-137-24-273

Abstract: With the recent development of LED lighting systems for plant cultivation, the use of vertical farming under controlled conditions is attracting increased attention. This study investigated the impact of a number of LED light spectra (red, blue, green and white) on the growth, development and essential oil content of lemon balm (*Melissa officinalis*), a herb and pharmaceutical plant species used across the world. White light and red-rich light spectra gave the best outputs in terms of impact on the growth and yield. For blue-rich spectra, the development and yield was lower despite having a significant impact on the photosynthesis activity, including Fv/Fm and NDVI values. For the blue-rich spectra, a peak wavelength of 450 mn was better than that of 435 nm. The results have practical value in terms of increased yield and the reduction of electricity consumption under controlled environmental conditions for the commercial production of lemon balm.

Keywords: vertical farming; 435 nm; 450 nm; white light; light quality; growth; chemical profile

Citation: Rihan, H.Z.; Aljafer, N.; Jbara, M.; McCallum, L.; Lengger, S.; Fuller, M.P. The Impact of LED Lighting Spectra in a Plant Factory on the Growth, Physiological Traits and Essential Oil Content of Lemon Balm (*Melissa officinalis*). *Plants* **2022**, *11*, 342. https://doi.org/10.3390/plants11030342

Academic Editor: Fermin Morales

Received: 29 July 2021
Accepted: 26 January 2022
Published: 27 January 2022

1. Introduction

Lemon balm belongs to the Mint family (*Lamiaceae*) and grows widely in central and southern Europe and in Asia minor. It is cultivated globally because of its culinary and medicinal properties [1]. It has important applications as a herbal treatment for stress and anxiety, and has antioxidant properties that are of use in pharmaceutical applications [2] and is used in perfumes, cosmetics, tea and food products [3]. It also has antibacterial properties and a sedative impact, and these are attributed to its flavonoid and essential oil content [4,5]. The chemical composition of lemon balm leaves also include polyphenolic compounds, such as trimeric compounds, rosmarinic acid and other flavonoids. Its leaves are used in raw form as a salad vegetable in various parts of the world [6].

Light is one of the main factors influencing the physiology, growth, development and chemical composition of plants [7]. The major impact of light in plants is on photosynthesis which utilises Photosynthetically Active Radiation (PAR) comprising wavelengths of light between 400–700 nm However, plants do not respond uniformly to all wavelengths of PAR and red (600–700 nm) and blue wavelengths (420–460 nm) are the most effective at driving photosynthesis due to the absorption capacity of the light absorbing pigments chlorophyll a and b. Other wavebands play a crucial role in photo-morphological development, especially in the far-red region (above 700 nm) and some might cause harm to plant cell DNA (below 400 nm, for example) [8].

Horticultural Light Emitting Diodes (LEDs) modules have been recently developed as artificial or supplementary grow lights [9]. They have potential for use as supplementary lighting in glasshouses and sole-source lighting options in plant factory systems, where plants are grown indoors under controlled environmental conditions [10]. LEDs have many positive features, such as linear photon output, durability and long operating lifespan, as well as a capacity for construction in large arrays that produce high PAR suitable for plant growth and development. Furthermore, LED modules emit less heat than traditional lighting systems such as High-Pressure Sodium, Halide and Fluorescent tubes [11–16]. More importantly, spectral specificity can be introduced through the design of the LED array, utilising a mixture of LEDs with different wavelengths and this can be managed through the appropriate control systems [17]. This, in turn, has a high research and commercial application, due to the fact that plant species respond differently to various wavelengths, owing to specific differences in their photoreceptors [8,18].

The impact of LEDs on the growth, shape, yield and edible quality parameters of several plant species has been the subject of an increasing amount of recent research [19,20]. Moreover, a significant amount of research has confirmed the impact of LEDs on chemical composition, such as vitamin C content, soluble sugar [21], chlorophyll content [22] and the protein level and anti-oxidant activity of several plant species [23].

That said, a certain amount of research on the impact of LEDs and light wavelength on lemon balm has been previously reported [7,24]. The authors of [7] compared the use of florescent lamps (FL) with the use of white LED lights (Philips LEDs) and they reported that light sources did not have a significant impact on the growth and yield of lemon balm plants, but these plants were featured by a higher net photosynthesis value when grown under FL lamps as compared to LEDs. It was also reported that lemon balm's chemical composition was significantly affected by the lighting conditions; for example, lemon balm had a higher content of macro- and micronutrients when they were grown under LEDs compared with fluorescent lamps [24]. Despite the published research about the impact of light conditions on the growth and yield of lemon balm, a better understanding of the impact of wavelength on the physiology, growth, yield, essential oil content and quality is still needed. The aim of this study, therefore, is to investigate the impact of wide range of light spectra including various combinations of blue, red and green on the physiological and chemical traits of lemon balm. One of its main objectives is also the investigation of the impact of blue-light wavelength sources on the physiology, growth and quality of lemon balm, since our recently published research showed that this could have a great impact on both growth and quality of some plant species, such as basil [17].

2. Material and Methods

2.1. Plant Material

Lemon balm (*Melissa officinalis*) seeds were obtained from CN seeds (CN Seeds, Pymoor, Ely, Cambridgeshire, UK). Seeds were sown and germinated in Rockwool cubes (36 mm) under dark conditions and 22 ± 2 °C for 10 days, and were then transferred to an Ebb and Flow hydroponic system in the Plant Factory facility at the University of Plymouth. The Plant Factory facility is a converted insulated greenhouse, in which external light has been excluded and a multi-tier hydroponic growing system, consisting of Ebb and Flow trays with interchangeable LED light units, has been installed. The Plant Factory system is divided into several multi-shelf hydroponic units, each consisting of three tiers. The distance between tiers is 50 cm. Temperature and humidity were monitored using Gemini data loggers (Tinytag Plus (part No GP-1590)) and an instantaneous thermometer (Fisher Scientific, Loughborough, UK) at 23 ± 2 °C and humidity at 65 ± 5%. The dark/light period was set to 8/16 h. Five lighting treatments were designed and applied using LuminiGrow LED lighting systems (LuminiGrow, Shenzhen, China). The light treatments included several combinations of blue (B), green (G) and red (red) at different ratios described as follows: T1—white (B:G:R, 1:2.3:2); T2:—blue-rich with blue peak at 450 nm (B:G:R, 1:0.02:0.8);

T3—blue-rich with blue at 450 nm + green (B:G:R, 1:0.07:0.65); T4—blue-rich with blue at 435 nm (B:G:R, 1:0.02:0.8); and T5—red-rich (B:G:R, 1:0.025:1.6) (Figures 1 and 2).

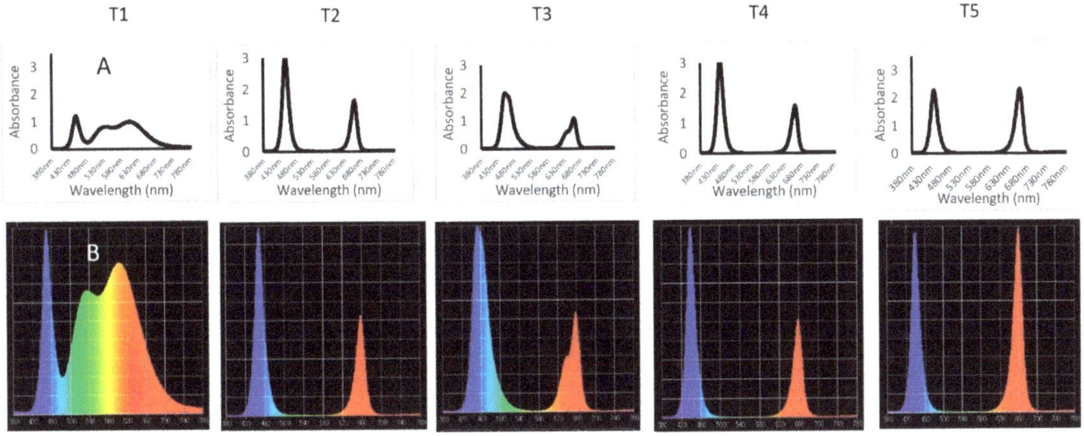

Figure 1. Spectra of the lighting treatments used, as measured by a UPRtek spectrophotometer: (**A**) the radiant density of the light spectrum intensity and (**B**) the relative light intensity.

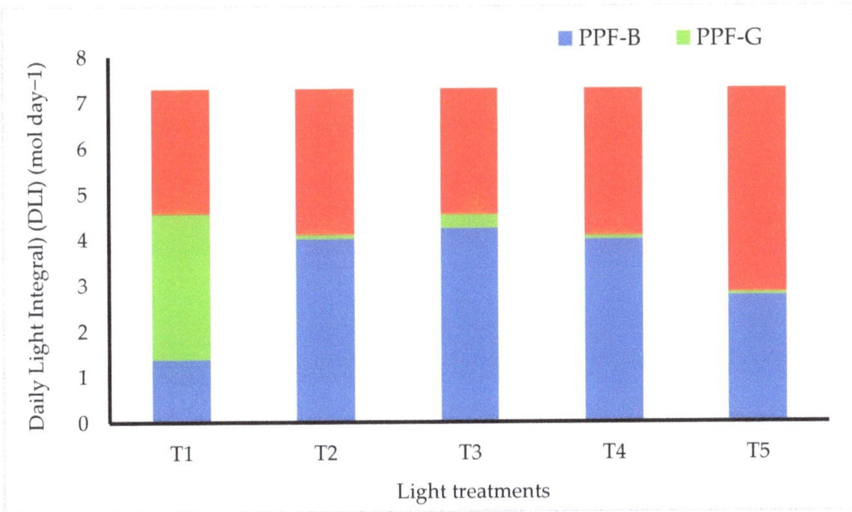

Figure 2. Daily light integral for different light spectra of the applied light treatments.

Light intensity from the LED lighting treatments was measured using a UPRtek spectrophotometer (UPRtek MK350N premium Standalone handheld spectral light meter, Taiwan) and adjusted to deliver 125 ± 10 μmol m^{-2} s^{-1}. Daily light integral to the LED light treatments were calculated at 7.2 mol day^{-1} (Figure 2). The emitted light spectra of the lighting treatments were measured (relative light intensity) using a UPRtek spectrophotometer and corrected to show the radiant density at each wavelength (Figure 1).

Growth and physiological responses of lemon balm to the lighting treatments were measured at the harvest stage (64 days from sowing). Growth/yield measurements including plant height (cm) (n = 8) and leaf area (LA cm^2) (n = 8) were made, using a leaf area image analyser HITACHI KP-D40 colour digital camera with a lightbox and WinDias

1.5 software (Delta-T Devices Ltd., Cambridge, UK). Fresh weight (FW) (n = 4) and dry weight (DW) (g) (n = 4) were measured after removing the root system, using a Fisher Scientific SG-402 laboratory balance. For dry weight, plants were dried at 60 °C for 96 h [25].

2.2. Chlorophyll Fluorescence Imaging System (Fv/Fm and NDVI)

Fluorescence image acquisition was performed with a PSI open FluorCam FC 800-O (PSI (Photon Systems Instruments), Drasov, Czech Republic) applying a protocol derived from Méline et al. [26]. The system has four individual LED panels in two pairs. The first pair of LED panels provides an orange actinic light with an intensity up to 400 μmol m^{-2} s^{-1}, and a wavelength of approximately 620 nm. A second pair of LED panels gives a saturating pulse with an intensity up to 3000 μmol m^{-2} s^{-1}, with a blue wavelength around 455 nm. The system sensor is a CCD camera, which has a pixel resolution of 512 by 512 and a 12-bit dynamic. When photosynthetic yield is at zero, fluorescence emission reached a maximum (Fm). Fv, a variable fluorescence, defined as the difference Fm − F0. Fv and Fm, are used to calculate the maximum quantum yield of QY max = Fv/Fm. This was measured after 20 min of dark adaptation. Another important vegetative index called "normalized difference vegetation index" (NDVI), is based on the spectral reflectance of plants in the near infrared region (λ = 700–1300 nm) and the visible red range (λ = 550–700 nm) of the electromagnetic spectrum. Dark adaptation is 20 min and calculation of NDVI is:

$$\mathrm{NDVI} = (\mathrm{NIR} - \mathrm{Red})/(\mathrm{NIR} + \mathrm{Red})$$

2.3. Chlorophyll Content

Chlorophyll content was evaluated, using the method described by [27]. Plant tissue (leaves) (0.2 g) was ground with 10 mL 80% acetone. The final volume was made up to 10 mL with 80% acetone and then centrifuged for 3 min. Absorbance was measured against an 80% acetone blank. Supernatant (2 mL) was placed in a cuvette and the absorbance was measured at 663.6 ($A_{663.6}$) and 646.6 ($A_{646.6}$), using a Jenway 7315 (Staffordshire, UK). The formulae are based on the absorbance maxima of each pigment and are dependent on the solvent used. The formulae for samples dissolved in acetone are as follows:

$$\mathrm{Ca} = 12.25\ A_{663.6} - 2.55\ A_{646.6}$$

$$\mathrm{Cb} = 20.31\ A_{646.6} - 4.91\ A_{663.6}$$

$$\mathrm{Total\ C} = 17.76\ A_{646.6} + 7.34\ A_{663.6}$$

where Ca: chlorophyll A, Cb: chlorophyll B, Total C: total chlorophyll.

The values obtained were converted to estimate the chlorophyll content per gram of fresh weight, following the procedure described by [27].

2.4. Essential Oil Analysis

Leaves were collected and dried, then 10 g were ground in a mortar and pestle. The essential oil was extracted employing the Soxhlet method, using absolute ethanol (Thermo Fisher Scientific, Loughborough, UK) as a solvent [28]. Through this method, 10 g of dry lemon balm was extracted for 4 h. Once the extraction was complete, essential oil was separated from the solvent, using a BÜCHI R-124 Rotary Evaporator System (BUCHI UK Ltd., Suffolk, UK). The essential oils were then collected in a vial. The vial was weighed before and after the extraction to calculate the quantity of the essential oil obtained.

2.5. Statistical Analysis

The main experiment consisted of 5 lighting treatments with 4 replicates, each consisting of 15 plants. Treatments were randomised at each replication. Results were presented as means ± standard error (S.E.). All data were subjected to analysis of variance (ANOVA) using Minitab software (version 19). Tukey' post hoc test was used for determination of significant differences between the treatments. Comparisons of means were made using the least significant difference (LSD) test at a 95% level of probability.

3. Results

3.1. Growth/Yield Responses to LED Light Treatments

Although all lighting treatments produced plants of an acceptable commercial quality, there was a significant impact of different spectra on various growth and yield parameters.

White light (T1), which is the only spectrum that included a high level of green wavelengths in addition to red and blue, had a significant impact of the average fresh weight ($p = 0.004$), average dry weight ($p = 0.018$), plant height ($p \leq 0.001$) and internodes ($p \leq 0.001$) in comparison with other light treatments (Figure 3A–C,F). Red-rich treatment (T5) also seemed to have a positive impact on plant height and internodes in comparison with other treatments, apart from the white light treatment (Figure 3C,F).

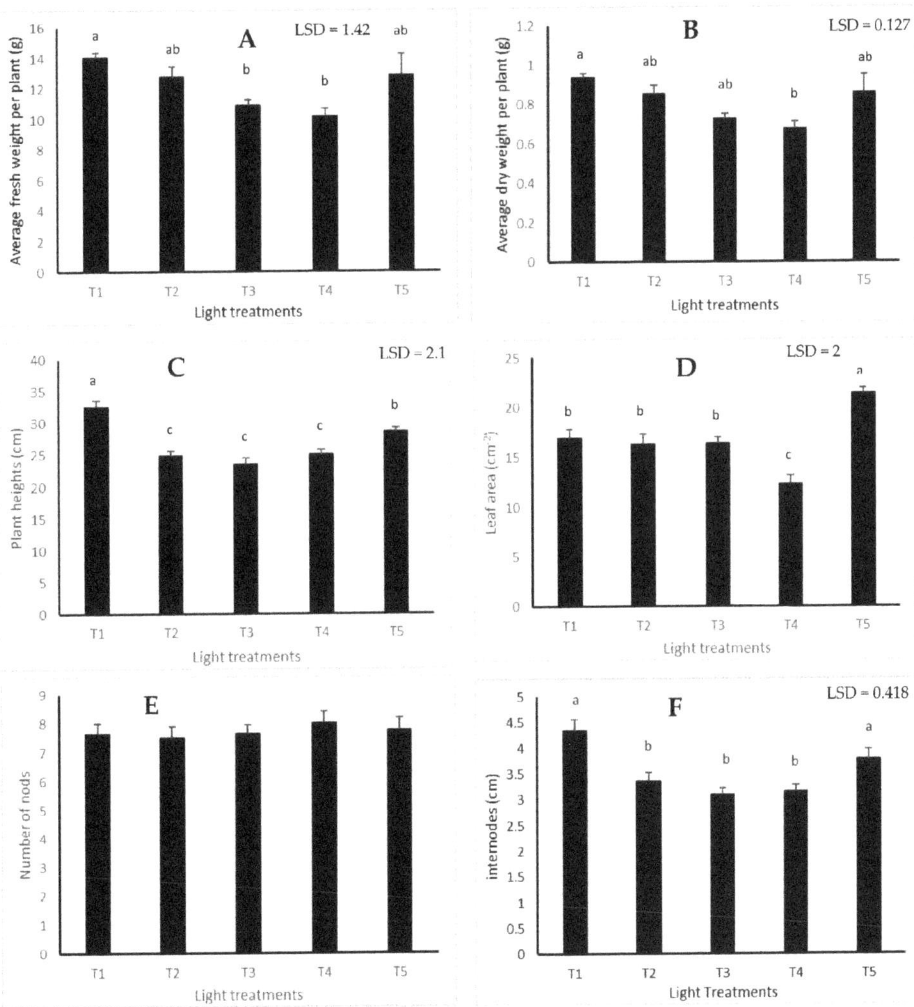

Figure 3. The effect of light treatment on growth traits: (**A**) fresh weight (g); (**B**) dry weight (g); (**C**) plant heights (cm); (**D**) leaf area; (**E**) number of nodes; (**F**) internodes (cm) of lemon balm measured

at the harvest stage (64 days from sowing) (Means denoted by a different letter indicate significant differences between treatments ($p < 0.05$)).

Red-rich light treatment (T5) had a significant positive impact on leaf area compared with other light spectra ($p \leq 0.001$). Light spectrum did not significantly impact the number nodes ($p = 0.918$) (Figure 3E).

With regards to the impact of blue light wavelength (T2 and T4), using blue light at 450 nm (T2) produced larger plants (fresh and dry weights) and bigger leaf area compared to the use of 435 nm (T4) (Figure 3A,B,D respectively). However, no significant impact of the blue source was observed on plant height (Figure 3C).

3.2. Physiological and Chemical Responses to LED Light Treatments

3.2.1. Fv/Fm Ratio and NDVI Indicators

Light treatments had a significant impact on Fv/Fm value ($p = 0.016$) and NDVI ($p = 0.024$) (Figure 4). Blue-rich spectrum with 435 nm used as a source of blue light had a significant impact on both Fv/Fm and NDVI. White spectrum (T1) and blue/red treatment that included some green in the spectrum combination (T3) had a negative impact on both Fv/Fm and NDVI (Figure 4).

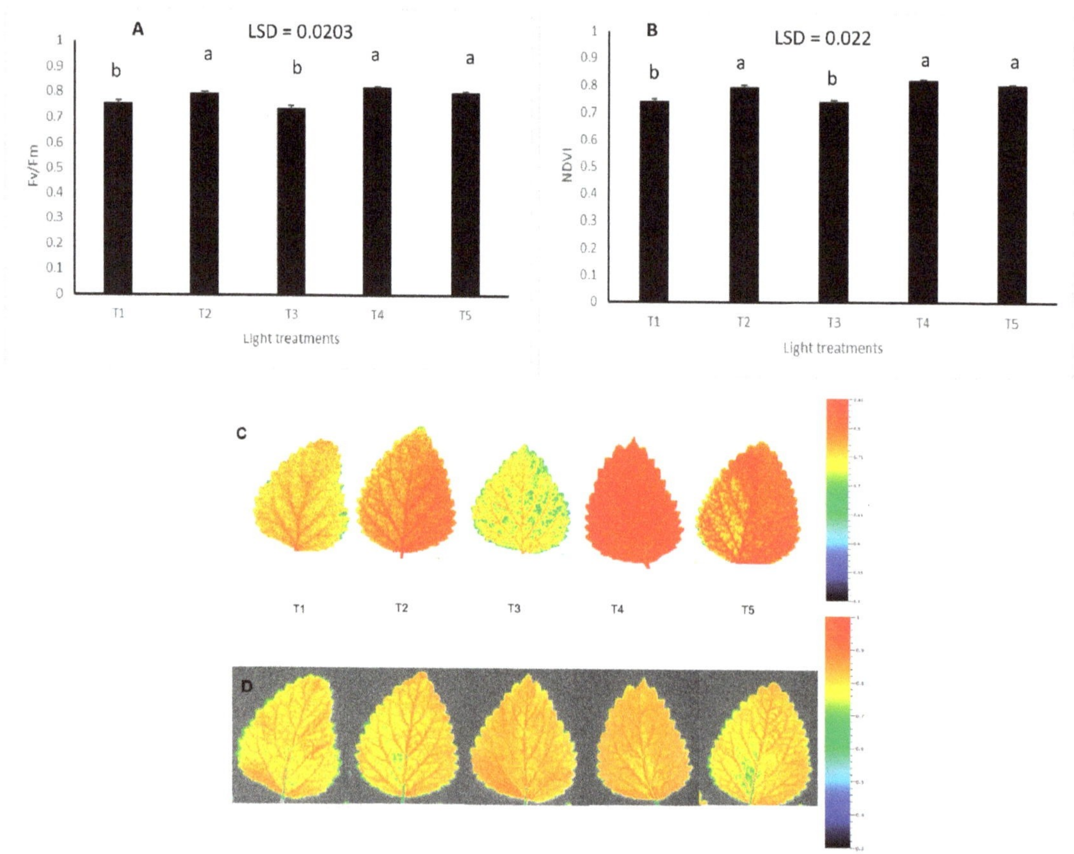

Figure 4. The effect of light treatments on Fv/Fm (**A**,**C**) and NDVI (**B**,**D**) (Means denoted by a different letter indicate significant differences between treatments ($p < 0.05$)).

3.2.2. Chlorophyll Content

There were only small and non-significant differences in chlorophyll content among the lighting treatments either for chlorophyll A, chlorophyll B or total chlorophyll ($p = 0.227$, $p = 0.620$ and $p = 0.315$ respectively) (Figure 5).

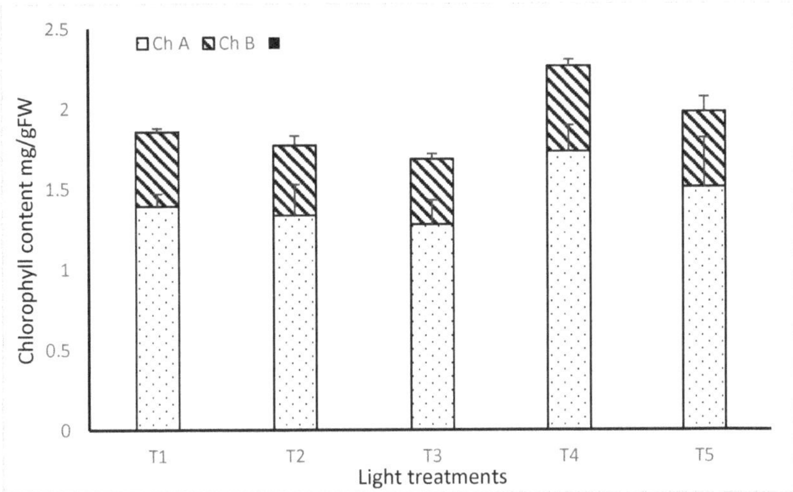

Figure 5. The effect of light treatment on chlorophyll content of lemon balm.

3.3. The Impact of Light Treatment on the Essential Oil Content in Lemon Balm

Light treatment had a significant impact on essential oil content per plant of lemon balm. White light treatment had a significant impact on the essential oil yield of lemon balm in comparison with other light treatments ($p \leq 0.001$) (Figure 6). Moreover, Blue-rich with blue at 435 nm (T4) had a negative impact on the essential oil content.

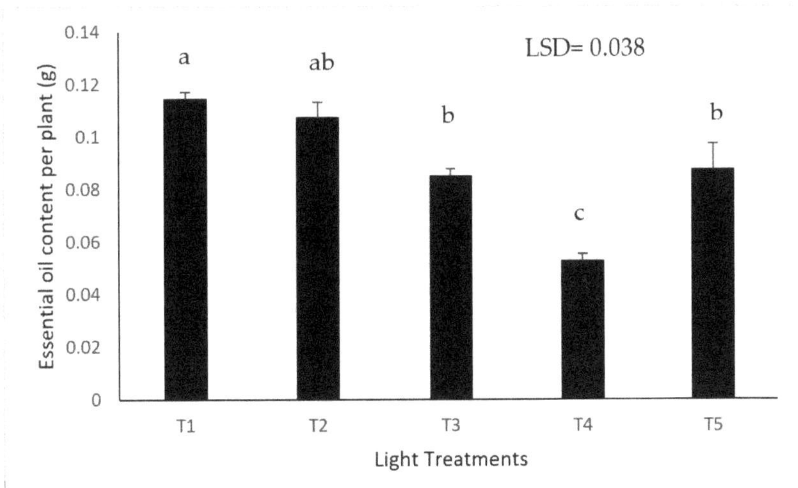

Figure 6. The effect of light treatment on essential oil content per lemon balm plant (Means denoted by different letters indicate significant differences between treatments ($p < 0.05$)).

4. Discussion

Melissa officinalis (lemon balm) is an important source of active chemicals, such as triterpenes, flavonols, phenolic acid and many other important pharmaceutical compounds [29]. Similarly to other pharmaceutical plant species, the growth, yield and chemical compositions of these species are affected by environmental factors when grown under open-field conditions. However, with the recent fast development of LED grow lighting systems and the increasing efficacy of these systems, growing pharmaceutical plants, such as lemon balm, vertically under controlled environmental conditions is now viable and has potential commercial value. Vertical farming (plant factory) is a novel plant-production system that allows local production of high-quality pharmaceutical plant species [30]. In the plant factory system, LEDs are used as the sole source of lighting and provide a unique tool for promoting growth, yield and quality. However, plant species respond differently to lighting conditions and therefore it is crucial to vary light spectra and intensity to suit the requirements of individual plant species [4,17,31].

In terms of lemon balm, white (50% cool white + 50% warm light) light improved the growth traits, including fresh and dry weight, plant heights and internode spacing. This highlights the great impact of green spectrum on the growth of this plant species, since white light has a significant amount of green, which is higher than both the blue and the red spectra. This also could be due to the impact of other wavelengths such as orange, yellow, etc., which existed in the full spectrum of white and did not exist in the other described treatments in this research. This finding is in accord with other research indicating the positive impact of white light on the growth and development of plants, even by comparison with blue light added to red LEDs [32].

Kim et al. [20] reported that lettuce plants grown with spectra that included green light had better growth levels, including fresh and dry weights, than those grown with red/blue only. However, the current findings disagree with those of [10,17] which had indicated the positive impact on the development and growth of basil of focusing light in the red and blue regions. The observed differences could, however, be due to the differing responses to light of the two plant species.

The current finding showed a high positive impact of red-rich light spectra on the growth parameters of lemon balm, including fresh and dry weight, height and leaf area. Lin, Huang and Hsu [33] reported a significant positive impact of high level red light on the growth and development of green and purple basil plants. Red light is one of the essential components in lighting spectra for plant growth and red light alone is sufficient for normal plant growth and photosynthesis [34]. The current results are also in accord with what has been reported of the yield reduction associated with a high level of blue light in light spectra, a phenomenon that had been linked previously with lower internode length and smaller leaf area [32,35], and which was also observed in the current study. It was reported that red light matches the assimilation peak of the photoreceptors phytochrome and chlorophyll and that the combination of red–blue light for growing plants causes a greater improvement in the maximum photosynthetic rate than monochromatic light, as a consequence of the activation of cryptochromes, phytochromes and chlorophyll [36]. However, the current findings do not agree with what was reported by [4,17] concerning the significant impact of high-level blue light compared with red light in the light spectrum on the growth and development of basil. These conflicting results could be attributed to a difference in experimental conditions, such as the light intensity, the temperature and the plant species that was studied. One of the main challenges to the replicability of research results in LED lighting applications could be caused by the high variability of experimental setups [17,32]. For example, while 125 μmol m^{-2} s^{-1} PPFD was applied in the current research, high PPFD values, e.g., 300 μmol m^{-2} s^{-1}, were applied by [17].

The use of 435 nm as a source of blue light instead of 450 nm, which is widely used in commercial undertakings, had a negative impact on the yield parameter of lemon balm. This contrasts with what was reported by [17] regarding the significant positive impact of this wavelength on the growth and development of basil. However, a significant number of research studies have reported the role of blue at 450 nm on the growth and development of plants [32]. These differences could be due to the genetic, physiological differences between plant species. Further research on the impact of blue light variations on plants is necessary.

Light spectra did not have a significant impact on the chlorophyll content of lemon balm. It is possible that the level of chlorophyll was not affected by the light treatment because all the treatment contained a sufficient level of blue and red in the spectrum combinations. It has previously been reported that LED light supplying RB increased the total amount of chlorophyll in Chinese cabbage leaves, compared with the concentration of chlorophyll in plants treated with blue or red light only [37,38]. Chen et al. [35] found that the chlorophyll content of lettuce leaves was higher when plants were grown under a mixture of red and blue spectra, compared to growth with blue or red light only.

The impact of light spectra on the photosynthesis activity of lemon balm was evaluated using a Chlorophyll Fluorescence Imaging system. Chlorophyll fluorescence imaging is an extremely important technique for the non-invasive study of photosynthesis dynamics in intact plants, algae and in cyanobacteria for the measurement of chlorophyll fluorescence kinetics. This device/technique was used to calculate the maximum quantum yield of QY max = Fv/Fm. An interesting finding of the current study was that blue light at 435 nm has a significant positive impact on both Fv/Fm and NDVI indicators compared to other light treatments. Moreover, the use of 435 nm as a source of blue has a significant impact on these indicators as compared to the same treatment with 450 nm used as a source of blue. This finding agrees with that of Rihan et al. (2020) on the significant impact of 435 nm wavelength compared with 450 nm wavelength in terms of its effects on the photosynthesis activity of basil. The 435 nm treatment had a positive impact on the stimulation of PS I in the photosynthesis process in *Cyanobacteria Bacteria* and *Arabidopsis thaliana* [39]. This could explain the significant increase in the Chlorophyll Fluorescence Rate (Fv/Fm) and the NDVI indicator observed in the current research. A fluorescence spectral analysis showed that Chamomile pollen reaches a peak in a blue light region of 435 nm [40]. However, in the current study, the significant photosynthesis activities did not translate into an improvement in the growth rate of lemon balm. There could be several reasons for this, including differences in the experimental conditions, such as light intensity, temperature, etc. More research is needed for a further understanding of the conflicting findings with regard to the photosynthesis parameters and growth traits observed in this plant species.

Although no significant impact of light spectrum on the content of essential oil was observed, there was a clear negative impact of blue 435 nm on the essential oil content. However, further studies of the impact of light spectra on the quality and chemical composition of lemon balm oil are needed.

5. Conclusions

Between a wide range of light spectra, including white, red/blue in various ratios and blue at different wavelengths, the best results in terms of the impact of light spectra on growth and yield were obtained using white light (50% cool white + 50% warm white). This has a high practical application, as white light has wide commercial availability and is user friendly. Moreover, blue light sources seem to have a significant impact on the growth and physiology of lemon balm. While blue at 450 nm promoted growth and increased the yield, blue at 435 nm had a significant impact on the photosynthesis activities.

Author Contributions: H.Z.R.: Conceptualization, Methodology, Supervision, Writing—review & editing. N.A.: Investigation, Writing—original draft. M.J.: Investigation. L.M.: Investigation. S.L.: Investigation. M.P.F.: Conceptualization, Methodology, Supervision, Validation, Writing—review & editing, Project administration. All authors have read and agreed to the published version of the manuscript.

Funding: This research was partially funded by the European Regional Development Fund for Cornwall, UKunder an AgriTech research grant and the Seale-Hayne Educational Trust (grant RR204307).

Institutional Review Board Statement: Not applicable.

Informed Consent Statement: Not applicable.

Data Availability Statement: Data is contained within the article.

Acknowledgments: Technical and scientific assistance are gratefully acknowledged from Demelza Carne (Plymouth University).

Conflicts of Interest: The authors declare no conflict of interest. The funders had no role in the design of the study; in the collection, analyses, or interpretation of data; in the writing of the manuscript; or in the decision to publish the results.

References

1. Dastmalchi, K.; Damien Dorman, H.J.; Oinonen, P.P.; Darwis, Y.; Laakso, I.; Hiltunen, R. Chemical composition and in vitro antioxidative activity of a lemon balm (*Melissa officinalis* L.) extract. *LWT-Food Sci. Technol.* **2008**, *41*, 391–400. [CrossRef]
2. Asadi, N.; Ramezanzadeh, M.; Bahlakeh, G.; Ramezanzadeh, B. Theoretical MD/DFT computer explorations and surface-electrochemical investigations of the zinc/iron metal cations interactions with highly active molecules from Lemon balm extract toward the steel corrosion retardation in saline solution. *J. Mol. Liq.* **2020**, *310*, 113220. [CrossRef]
3. Ahmadi, T.; Shabani, L.; Sabzalian, M.R. Improvement in drought tolerance of lemon balm, *Melissa officinalis* L. under the pre-treatment of LED lighting. *Plant Physiol. Biochem.* **2019**, *139*, 548–557. [CrossRef] [PubMed]
4. Mohammadi, G.; Rafiee, G.; El Basuini, M.F.; Van Doan, H.; Ahmed, H.A.; Dawood, M.A.; Abdel-Latif, H.M. Oregano (*Origanum vulgare*), St John's-wort (*Hypericum perforatum*), and lemon balm (*Melissa officinalis*) extracts improved the growth rate, antioxidative, and immunological responses in Nile tilapia (*Oreochromis niloticus*) infected with *Aeromonas hydrophila*. *Aquac. Rep.* **2020**, *18*, 100445. [CrossRef]
5. Bilen, S.; Altief, T.A.S.; Özdemir, K.Y.; Salem, M.O.A.; Terzi, E.; Güney, K. Effect of lemon balm (*Melissa officinalis*) extract on growth performance, digestive and antioxidant enzyme activities, and immune responses in rainbow trout (*Oncorhynchus mykiss*). *Fish Physiol. Biochem.* **2020**, *46*, 471–481. [CrossRef]
6. Patora, J.; Klimek, B. Flavonoids from lemon balm (*Melissa officinalis* L., Lamiaceae). *Acta Pol. Pharm.* **2002**, *59*, 139–144.
7. Frąszczak, B.; Golcz, A.; Zawirska-Wojtasiak, R.; Janowska, B. Growth rate of sweet basil and lemon balm plants grown under fluorescent lamps and LED modules. *Acta Sci. Pol. Hortorum Cultus* **2014**, *13*, 3–13.
8. Sims, D.A.; Gamon, J.A. Relationships between leaf pigment content and spectral reflectance across a wide range of species, leaf structures and developmental stages. *Remote Sens. Environ.* **2002**, *81*, 337–354. [CrossRef]
9. Losev, O.; Biard, J.; Holonyak, N. *Light-Emitting Diode*; S.Hirzel Verlag Publisher: Leipzig, Germany, 2016; pp. 1–25.
10. Aldarkazali, M.; Rihan, H.Z.; Carne, D.; Fuller, M.P. The growth and development of sweet basil (*Ocimum basilicum*) and bush basil (*Ocimum minimum*) grown under three light regimes in a controlled environment. *Agronomy* **2019**, *9*, 743. [CrossRef]
11. Green, M.A.; Zhao, J.; Wang, A.; Reece, P.J.; Gal, M. Efficient silicon light-emitting diodes. *Nature* **2001**, *412*, 805. [CrossRef]
12. Van Ieperen, W.; Trouwborst, G. The application of LEDs as assimilation light source in greenhouse horticulture: A simulation study. In Proceedings of the International Symposium on High Technology for Greenhouse System Management: Greensys, Naples, Italy, 4–6 October 2007; pp. 1407–1414.
13. Tonzani, S. Lighting technology: Time to change the bulb. *Nat. News* **2009**, *459*, 312–314. [CrossRef] [PubMed]
14. Singh, D.; Basu, C.; Meinhardt-Wollweber, M.; Roth, B. LEDs for energy efficient greenhouse lighting. *Renew. Sustain. Energy Rev.* **2015**, *49*, 139–147. [CrossRef]
15. Bosma, R.H.; Lacambra, L.; Landstra, Y.; Perini, C.; Poulie, J.; Schwaner, M.J.; Yin, Y. The financial feasibility of producing fish and vegetables through aquaponics. *Aquac. Eng.* **2017**, *78*, 146–154. [CrossRef]
16. Dickinson, J. *Inventory of New York City Greenhouse Gas Emissions*; DIANE Publishing: Collingdale, PA, USA, 2007.
17. Rihan, H.Z.; Aldarkazali, M.; Mohamed, S.J.; McMulkin, N.B.; Jbara, M.H.; Fuller, M.P. A Novel New Light Recipe Significantly Increases the Growth and Yield of Sweet Basil (*Ocimum basilicum*) Grown in Plant Factory System. *Agronomy* **2020**, *10*, 934. [CrossRef]
18. Brown, C.S.; Schuerger, A.C.; Sager, J.C. Growth and photomorphogenesis of pepper plants under red light-emitting diodes with supplemental blue or far-red lighting. *J. Am. Soc. Hortic. Sci.* **1995**, *120*, 808–813. [CrossRef]
19. Lin, K.-H.; Huang, M.-Y.; Huang, W.-D.; Hsu, M.-H.; Yang, Z.-W.; Yang, C.-M. The effects of red, blue, and white light-emitting diodes on the growth, development, and edible quality of hydroponically grown lettuce (*Lactuca sativa* L. var. *capitata*). *Sci. Hortic.* **2013**, *150*, 86–91. [CrossRef]
20. Kim, H.-H.; Goins, G.D.; Wheeler, R.M.; Sager, J.C. Green-light supplementation for enhanced lettuce growth under red-and blue-light-emitting diodes. *HortScience* **2004**, *39*, 1617–1622. [CrossRef]
21. Ohashi-Kaneko, K.; Takase, M.; Kon, N.; Fujiwara, K.; Kurata, K. Effect of light quality on growth and vegetable quality in leaf lettuce, spinach and komatsuna. *Environ. Control Biol.* **2007**, *45*, 189–198. [CrossRef]
22. Kim, K.; Kook, H.-S.; Jang, Y.-J.; Lee, W.-H.; Kamala-Kannan, S.; Chae, J.-C.; Lee, K.-J. The effect of blue-light-emitting diodes on antioxidant properties and resistance to Botrytis cinerea in tomato. *J. Plant Pathol. Microbiol.* **2013**, *4*. [CrossRef]

23. Wu, M.-C.; Hou, C.-Y.; Jiang, C.-M.; Wang, Y.-T.; Wang, C.-Y.; Chen, H.-H.; Chang, H.-M. A novel approach of LED light radiation improves the antioxidant activity of pea seedlings. *Food Chem.* **2007**, *101*, 1753–1758. [CrossRef]
24. Frąszczak, B.; Gąsecka, M.; Golcz, A.; Zawirska-Wojtasiak, R. The chemical composition of lemon balm and basil plants grown under different light conditions. *Acta Sci. Pol. Hortorum Cultus* **2015**, *14*, 93–104.
25. Saha, S.; Monroe, A.; Day, M.R. Growth, yield, plant quality and nutrition of basil (*Ocimum basilicum* L.) under soilless agricultural systems. *Ann. Agric. Sci.* **2016**, *61*, 181–186. [CrossRef]
26. Méline, V.; Brin, C.; Lebreton, G.; Ledroit, L.; Sochard, D.; Hunault, G.; Boureau, T.; Belin, E. A Computation Method Based on the Combination of Chlorophyll Fluorescence Parameters to Improve the Discrimination of Visually Similar Phenotypes Induced by Bacterial Virulence Factors. *Front. Plant Sci.* **2020**, *11*, 213. [CrossRef] [PubMed]
27. Porra, R.J. The chequered history of the development and use of simultaneous equations for the accurate determination of chlorophylls a and b. *Photosynth. Res.* **2002**, *73*, 149–156. [CrossRef] [PubMed]
28. Pallado, P.; Tassinato, G.; D'Alpaos, M.; Traldi, P. Gas chromatography/mass spectrometry in aroma chemistry: A comparison of essential oils and flavours extracted by classical and supercritical techniques. *Rapid Commun. Mass Spectrom.* **1997**, *11*, 1335–1341. [CrossRef]
29. Heshmati, J.; Morvaridzadeh, M.; Sepidarkish, M.; Fazelian, S.; Rahimlou, M.; Omidi, A.; Palmowski, A.; Asadi, A.; Shidfar, F. Effects of *Melissa officinalis* (Lemon Balm) on cardio-metabolic outcomes: A systematic review and meta-analysis. *Phytother. Res.* **2020**, *34*, 3113–3123. [CrossRef] [PubMed]
30. SharathKumar, M.; Heuvelink, E.; Marcelis, L.F.M. Vertical Farming: Moving from Genetic to Environmental Modification. *Trends Plant Sci.* **2020**, *25*, 724–727. [CrossRef]
31. Liu, J.; Osbourn, A.; Ma, P. MYB Transcription Factors as Regulators of Phenylpropanoid Metabolism in Plants. *Mol. Plant* **2015**, *8*, 689–708. [CrossRef]
32. Massa, G.D.; Kim, H.-H.; Wheeler, R.M.; Mitchell, C.A. Plant productivity in response to LED lighting. *HortScience* **2008**, *43*, 1951–1956. [CrossRef]
33. Lin, K.-H.; Huang, M.-Y.; Hsu, M.-H. Morphological and physiological response in green and purple basil plants (*Ocimum basilicum*) under different proportions of red, green, and blue LED lightings. *Sci. Hortic.* **2021**, *275*, 109677. [CrossRef]
34. Olle, M.; Viršile, A. The effects of light-emitting diode lighting on greenhouse plant growth and quality. *Agric. Food Sci.* **2013**, *22*, 223–234. [CrossRef]
35. Chen, X.-L.; Guo, W.-Z.; Xue, X.-Z.; Wang, L.-C.; Qiao, X.-J. Growth and quality responses of 'Green Oak Leaf'lettuce as affected by monochromic or mixed radiation provided by fluorescent lamp (FL) and light-emitting diode (LED). *Sci. Hortic.* **2014**, *172*, 168–175. [CrossRef]
36. Darko, E.; Heydarizadeh, P.; Schoefs, B.; Sabzalian, M.R. Photosynthesis under artificial light: The shift in primary and secondary metabolism. *Philos. Trans. R. Soc. B Biol. Sci.* **2014**, *369*, 20130243. [CrossRef]
37. Pennisi, G.; Blasioli, S.; Cellini, A.; Maia, L.; Crepaldi, A.; Braschi, I.; Spinelli, F.; Nicola, S.; Fernandez, J.A.; Stanghellini, C.; et al. Unraveling the Role of Red: Blue LED Lights on Resource Use Efficiency and Nutritional Properties of Indoor Grown Sweet Basil. *Front. Plant Sci.* **2019**, *10*, 305. [CrossRef] [PubMed]
38. Fan, X.; Zang, J.; Xu, Z.; Guo, S.; Jiao, X.; Liu, X.; Gao, Y. Effects of different light quality on growth, chlorophyll concentration and chlorophyll biosynthesis precursors of non-heading Chinese cabbage (*Brassica campestris* L.). *Acta Physiol. Plant.* **2013**, *35*, 2721–2726. [CrossRef]
39. Lamb, J.J.; Røkke, G.; Hohmann-Marriott, M.F. Chlorophyll fluorescence emission spectroscopy of oxygenic organisms at 77 K. *Photosynthetica* **2018**, *56*, 105–124. [CrossRef]
40. Roshchina, V.V. *Fluorescing World of Plant Secreting Cells*; Science Publishers: New York, NY, USA, 2008.

plants

MDPI

Article

Red Light and Glucose Enhance Cytokinin-Mediated Bud Initial Formation in *Physcomitrium patens*

Durga Prasad Biswal [1,2] and Kishore Chandra Sekhar Panigrahi [1,2,*]

1 School of Biological Sciences, National Institute of Science Education and Research (NISER), Bhubaneswar 752050, Odisha, India; dpbiswal@niser.ac.in
2 Homi Bhabha National Institute (HBNI), Training School Complex, Anushakti Nagar, Mumbai 400094, Maharashtra, India
* Correspondence: panigrahi@niser.ac.in

Abstract: Growth and development of *Physcomitrium patens* is endogenously regulated by phytohormones such as auxin and cytokinin. Auxin induces the transition of chloronema to caulonema. This transition is also regulated by additional factors such as quantity and quality of light, carbon supply, and other phytohormones such as strigolactones and precursors of gibberrelic acid. On the other hand, cytokinins induce the formation of bud initials following caulonema differentiation. However, the influence of external factors such as light or nutrient supply on cytokinin-mediated bud initial formation has not been demonstrated in *Physcomitrium patens*. This study deals with the effect of light quality and nutrient supply on cytokinin-mediated bud initial formation. Bud initial formation has been observed in wild type plants in different light conditions such as white, red, and blue light in response to exogenously supplied cytokinin as well as glucose. In addition, budding assay has been demonstrated in the *cry1a* mutant of *Physcomitrium* in different light conditions. The results indicate that carbon supply and red light enhance the cytokinin response, while blue light inhibits this process in *Physcomitrium*.

Keywords: red light; blue light; glucose; bud initial; nutation; phytochrome; cryptochrome; cytokinin

Citation: Biswal, D.P.; Panigrahi, K.C.S. Red Light and Glucose Enhance Cytokinin-Mediated Bud Initial Formation in *Physcomitrium patens*. *Plants* **2022**, *11*, 707. https://doi.org/10.3390/plants11050707

Academic Editors: Hail Rihan and Mick Fuller

Received: 31 December 2021
Accepted: 7 February 2022
Published: 7 March 2022

Publisher's Note: MDPI stays neutral with regard to jurisdictional claims in published maps and institutional affiliations.

1. Introduction

In recent years the moss *Physcomitrium patens* (formerly *Physcomitrella patens*) has emerged as a major non-angiosperm plant model [1–3]. Mosses are considered to be among the earliest land plants. They predominantly exist as haploid gametophyte. The sporophyte is represented by a gametophyte-dependent spore-producing capsule for a shorter time. The gametophytic moss body has two developmental forms such as (i) protonema, a filamentous structure and (ii) gametophore, a miniature plant-like leafy structure. Moss spores on germination give rise to protonema. Protonema consists of chloronema: a chlorophyll-rich structure, with individual cells separated by a longitudinal cell wall; and caulonema: chlorophyll-deficient protonema cells with oblique cell wall. Chloronema are the first-formed cells upon spore germination. Caulonema arise from chloronema and they further give rise to bud initials or gametophore buds, which later produce the leafy gametophore [1]. In mosses these transitions are under the control of hormones. Classical plant hormones such as auxin, cytokinins (CKs), and abscisic acid (ABA) are present in mosses including *Physcomitrium* [4–7]. Presence of true gibberellic acid (GA) has not been reported in mosses; however, intermediates of the GA biosynthesis pathway have been identified in *Physcomitrium* [8,9]. In addition, the presence of strigolactones (SLs) and ethylene have also been reported [10,11]. Auxin enhances the transition of chloronema to caulonema [12–16]. This transition is partly influenced by SLs and GA-biosynthesis intermediates [8,17]. On the other hand, CKs induce branching of caulonema and formation of bud initials in different mosses [18–24]. ABA induces brood cell or brachycyte formation in mosses [25]. Ethylene regulates submergence- and osmotic stress-related responses in *Physcomitrium* [11,26]

Hormone signaling pathways are primarily regulated in response to environmental signals such as light. Light and hormone signals interact to regulate multiple developmental responses across the plant kingdom from algae to flowering plants [27,28]. Sunlight is a mixture of different wavelengths of light. However, plants primarily rely on specific wavelengths of light such as blue, red, and far-red (FR) to regulate the developmental responses. Plants perceive the quality, quantity, direction, and duration of light by different photoreceptors such as phytochromes (PHYs), cryptochromes (CRYs), phototropins (PHOTs), ZEITLUPE (ZTL), and UV resistance locus-8 (UVR8) proteins [29]. PHYs sense the ratio of red light (RL) and FR light (R:FR) in the environment and enable the plants to avoid shade or low light conditions [30,31]. On the other hand, CRYs, PHOTs, and ZTLs perceive blue light (BL) and UVR8 protein responds to ultra violet-B (UV-B) light [32–35]. Most of these photoreceptors (except PHOTs) upon activation by light, interact with transcription factors and modulate gene expression. These photoreceptors regulate multiple growth and developmental responses such as seed germination, seedling de-etiolation, shoot and root development, plastid development, chloroplast relocation, flowering, shade avoidance, phototropism, circadian rhythm, and photoperiodism [36–38]. The early land plant model *Physcomitrium* possesses all of these photoreceptors except ZTL [39–43]. Light regulates spore germination, protonema branching, phototropism, polarotropism, and chloroplast movement in *Physcomitrium* [44–47].

In higher plants light and hormone signals interact to regulate seed germination, hypocotyl elongation, flowering, fruit and root development, phototropic responses, and shade avoidance [28,48]. Light signals have also been shown to influence hormonal response in mosses. Higher amount of light enhances the chloronema–caulonema transition in *Physcomitrium*, which is an auxin-mediated process. This is primarily a photosynthetic effect, where the high energy condition enhances the caulonema differentiation [49]. Not only light quantity but light quality also influences this differentiation. Caulonema formation is enhanced under RL in *Physcomitrium*. On the contrary, BL inhibits this response through suppression of auxin-signaling components [40]. It appears that RL and BL regulate auxin response in opposing manners to balance the developmental transition of *Physcomitrium* in the natural environment [27].

While the chloronema–caulonema transition is a two-dimensional (2D) division, formation of bud initials, which are the precursors of leafy gametophores, marks the initiation of three-dimensional (3D) growth in mosses. This is an important event in the acquisition of land habitat and evolution of plant form. Bud initial formation, which is a CK-mediated process, has been reported to be influenced by light quality. In *Pohlia nutans* bud initial formation is induced under RL, but suppressed in BL [50]. The inhibitory effect of BL is partly suppressed when the protonema are cultured in kinetin-supplemented medium [51] or in a mixture of BL and RL [50]. In a study, protonema of *Funaria* produced lesser number of bud initials in response to exogenously supplied CK under RL and BL as compared to white light (WL). Interestingly, the protonema side branches showed curling under RL (nutations) [52], which may be the early signs of bud initial formation. Protonema, cultured in glucose- and CK-supplemented medium under dark, produced bud initials upon exposure to regular pulses of RL. This enhancing effect of RL was partly inhibited upon exposure of FR [52]. These facts indicate that RL enhances the bud initial formation, while BL and FR light inhibit the same in *Funaria*. In *Physcomitrium*, growth responses of protonema and gametophores are regulated by PHYs [44], however the role of light in regulating bud initial formation has not been demonstrated clearly. *Physcomitrium* has seven copies of PHYs such as PpPHY1-PpPHY4 and PpPHY5a-PpPHY5c [39] as well as two copies of CRYs such as PpCRY1a and PpCRY1b [40]. PHYs regulate cytoplasmic events such as phototropism, polarotropism, and chloroplast movement in protonema [40,46]. CRYs regulate protonema branching and phototropism in *Physcomitrium*. The *cry* mutants of *Physcomitrium* produce more gametophores under WL. This indicates that CRYs most likely inhibit bud initial formation [40]. However, it is not clear how CRYs play a role in CK signaling. These mutants have altered sensitivity to auxin signals [40], but it is not known whether the sensitivity or biosynthesis of CKs is altered in these mutants.

Apart from light quality, nutrient or carbon supply (energy status) also affect the growth and development of mosses. In *Funaria*, glucose has been shown to enhance CK-induced bud initial formation and this process in the dark was shown to be accelerated in presence of both sucrose and kinetin, but not in presence of sucrose alone [21]. This indicates that bud initial formation is not dependent on carbon supply, but it enhances the effect of kinetin which induces bud induction in the absence of light [53]. Later it was shown that glucose accelerated protonema growth and bud initial formation in *Funaria* in low irradiance light but had no significant effect in high irradiance light. In fact, more number of bud initials were formed under high irradiance light compared to low irradiance light, irrespective of the presence or absence of glucose [52]. *Ceratodon purpureus* showed earlier bud initial formation in medium supplemented with both sucrose and kinetin compared to medium with kinetin alone [22]. In *Physcomitrium turbinatum* bud initials have been shown to develop, even when the requisite light was provided in a discontinuous manner (divided into different combinations of light and dark periods). It has been hypothesized that light is responsible for synthesis of a morphogenic substance, which is required in optimum amount for bud induction. This substance accumulates over time in a cumulative manner under discontinuous light and induces bud formation upon reaching the optimum level [54]. Sugars are suitable and logical candidates for this purpose [7]. In fact, bud formation was delayed when the protonema of *Physcomitrium turbinatum* were cultured in medium without sugar [54]. Most of the earlier studies conducted to elucidate CK action in mosses have used carbon sources such as glucose or sucrose. It has been postulated that carbon sources may have an enhancing effect on CK-mediated bud formation in mosses [7]. However, carbon sources may also have a negative effect on the growth and development of mosses. For example, glucose has been reported to enhance protonema growth (primarily caulonema) in *Bryum billarderi*, but it inhibits the bud initial formation in presence of CKs [55]. Interestingly glucose enhances protonema development and shoot growth in *Bryum argentium*, but inhibits these processes in *Atrichum undulatum* [56].

Light quality and energy supply have been demonstrated to have a significant effect on auxin-mediated protonema differentiation in *Physcomitrium* [40,49], but the influence of these factors on CK-mediated bud initial formation has not been demonstrated. In this study we evaluated bud initial formation in wild type (WT) protonema with respect to glucose and light quality (WL, BL, and RL) using growth chambers equipped with light emitting diodes (LEDs) as light sources. In addition, we also evaluated the budding response in the *cry1a* mutant of *Physcomitrium* under RL, BL, and mixture of RL and BL (BR) to establish the role of BL in CK-mediated budding. Results suggest that while RL and carbon supply promote gametophore bud formation, BL inhibits this process in *Physcomitrium*.

2. Materials and Methods

2.1. Culture and Growth Conditions

WT strain of *Physcomitrium patens* (Hedw.) Mitt. and *cry1a* mutant [40] were used as plant material. Protonema cultures were maintained in modified Knop solid medium (250 mg KH_2PO_4, 250 mg KCl, 250 mg $MgSO_4 \cdot 7H_2O$, 1000 mg $Ca(NO_3)_2 \cdot 4H_2O$, 12.5 mg $FeSO_4 \cdot 7H_2O$, 12 g agar in 1000 mL, pH 4.5) [57,58] overlaid with cellophane disk (80 mm diameter, type 325 P, AA Packaging, Preston, UK) in 9 cm petri dishes by weekly sub-culturing. The culture medium was supplemented with Hogland's A-Z trace element solution (1 mL/1000 mL Knop medium) (614 mg H_3BO_3, 389 mg $MnCl_2 \cdot 4H_2O$, 110 mg $Al_2(SO_4)_3 \cdot K_2SO_4 \cdot 24H_2O$, 55 mg $CoCl_2 \cdot 6H_2O$, 55 mg $CuSO_4 \cdot 5H_2O$, 55 mg $ZnSO_4 \cdot 7H_2O$, 28 mg KBr, 28 mg KI, 28 mg LiCl, 28 mg $SnCl_2 \cdot 2H_2O$, 25 mg $Na_2MoO_4 \cdot 2H_2O$, 59 mg $NiCl_2 \cdot 6H_2O$ in 1000 mL) [59]. One week old protonema tissues were ground with tissue homogenizer (Omni International, Kennesaw, GE, USA) and transferred to fresh Knop solid medium overlaid with a cellophane disk every week.

To observe the effect of light and glucose on bud initial formation, one week old protonema were ground and cultured in 100 mL liquid Knop medium (pH 4.5) in Erlenmeyer flasks capped with Silicosen® silicone sponge plugs (Hirschmann Laborgeräte, Eberstadt, Germany). The ground protonema were grown under constant light (70 $\mu mol \cdot m^{-2} \cdot s^{-1}$) for four days (22 °C) in a photo-incubator shaker (Innova 44 Incubator Shaker, New Brunswick, Eppendorf, Hamburg, Germany) at 220 rpm. The four-day old protonema were used to observe the effect of light and glucose on CK-mediated bud initial formation.

Kinetin solution (Sigma-Aldrich, Catalag No. K0753) was prepared by dissolving it in 0.1 M NaOH. Four-day old WT protonema (grown in liquid medium under constant light) were cultured in liquid Knop medium (pH 5.8) supplemented with kinetin (1 μM) and without kinetin (control) in different light conditions such as WL, BL, and RL both in the presence of glucose (1%) (Sigma-Aldrich, Catalog No. G8270) or without glucose for two days in a long day (LD) condition (16 hr light/8 hr dark). Individual colonies were then observed and photographed in an Olympus IX73 or Zeiss Stereo Discovery.V20 microscope. Images were analyzed by Image J for counting the buds.

To study the role of BL in bud initial formation, WT and *cry1a* protonema from four-day old suspension culture were grown in liquid Knop medium (pH 5.8) with glucose (1%) in the presence or absence of kinetin (1 μM) for two days under WL, RL, BL, and a mixture of BL and RL (BR light). Individual colonies were then observed, photographed and analyzed as described earlier. The role of BL was further verified by culturing the WT protonema in BR light in the absence of glucose with or without kinetin. WT protonema were also cultured in FR light (in glucose containing medium with and without kinetin) to analyze its effect on bud initial formation.

The results were analyzed by Kruskal–Wallis ANOVA using GraphPad Prism (Version 9) or OriginPro 2019. Post hoc analysis was carried out using Bonferroni's correction with a statistical significance of $p < 0.05$. The results were presented in median values using box plots.

2.2. Light Treatment

WL (50 $\mu mol \cdot m^{-2} \cdot s^{-1}$) was provided by plant growth chambers (Percival Scientific Inc. Perry, USA, Model No. CU36L6) equipped with fluorescent light, maintained at 22 °C. BL (30 $\mu mol \cdot m^{-2} \cdot s^{-1}$, 460 nm), RL (50 $\mu mol \cdot m^{-2} \cdot s^{-1}$, 670 nm), FR (50 $\mu mol \cdot m^{-2} \cdot s^{-1}$, 730 nm) and BR light were provided by LED chambers (Percival Scientific Inc. Model No. E-30LEDL3) maintained at 22 °C in a LD cycle. Light intensity and wavelength were measured by SpectraPen LM 510 (PSI Instruments, Drásov, Czech Republic).

2.3. Phytohormone Estimation

Seven-day-old protonema tissues cultured on cellophane disks were transferred to solid Knop medium (pH 5.8) supplemented with glucose and cultured for seven days in 16/8 hr light/dark cycle in WL. The tissue was harvested and snap-frozen in liquid nitrogen followed by grinding with liquid nitrogen. The ground plant material was then lyophilized (FreeZone 4.5, Labconco, Kansas City, MO, USA). *trans*-Zeatin (*t*Z) content was estimated according to Šimura et al. [60] (modified protocol). 25 mg of the lyophilized tissue was extracted in cold extraction buffer consisting of MeOH:H_2O:HCOOH (15:4:0.1) with 25 mg *trans*-[$^2H_5$] Zeatin as internal standard. *t*Z was then purified and quantified by liquid chromatography coupled with triple-quadrupole-trap MS/MS (QTRAP 6500+ LC-MS/MS, SCIEX, Framingham, MA, USA). The experiment was repeated three times and data were presented as mean ± SEM. Phytohormone estimation was done in the metabolomics facility of the National Institute of Plant Genome Research (NIPGR), New Delhi, India.

3. Results

3.1. Gametophore Bud Formation Is Increased in Red Light and in Presence of Glucose

Kinetin-induced bud initial formation was observed in WT protonema of *Physcomitrium* under WL, BL, and RL in the presence and absence of glucose.

Bud initial formation was not observed in the controls (no kinetin), neither in the presence nor the absence of glucose (Figure 1a–c,g–i and Figure 2(i)a–c,g–i,(ii)a–c,g–i). No bud initials or only insignificant amounts of bud initials were developed in protonema cultured in the medium lacking glucose, but supplemented with kinetin in all light conditions (Figures 1d–f and 2(i)d–f,(ii)d–f).

Gametophore buds were formed, when protonemata were cultured in the presence of both glucose and kinetin under all light conditions studied (Figure 2(i)j–l,(ii)j–l). Bud initial formation was observed to be much less in WL and BL. However, RL has a significant effect on bud initial formation and highest number of gametophores were formed under this light (Figure 3).

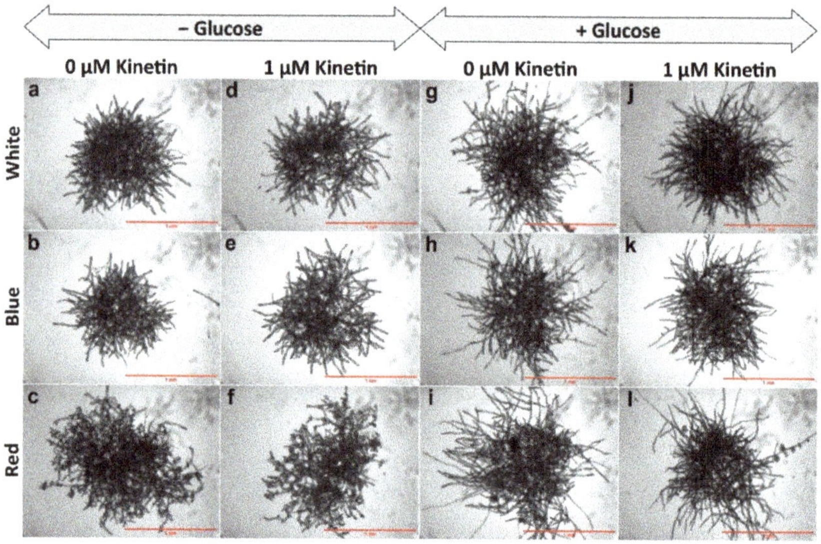

Figure 1. Morphology of protonema colonies under different light conditions. Protonema cultured in the presence of glucose show more growth (longer filaments) (**g**–**l**) compared to those in the absence of glucose (**a**–**f**). Protonema colonies cultured in red light in the absence of glucose do not show protruding filaments (**c**,**f**) as observed in other colonies, but show curling of filaments or nutations (peripheral regions). Scale Bar = 1 mm.

Figure 2. (**i**) Portions of protonema colonies showing the peripheral regions. Colonies under WL and BL show chloronema growth (green filaments). Colonies under RL show caulonema filaments (i and l) having less chlorophyll. Red triangles show the nutations and red arrows show the bud initials under RL. Scale Bar = 500 µm (0.5 mm). (**ii**) Portions of protonema colonies (magnified) showing presence or absence of nutations (red triangles) and bud initials (red arrows) under different light conditions. Nutations are present in RL in the absence of glucose, but bud initials are present in the presence of both glucose and kinetin. Scale Bar = 100 µm.

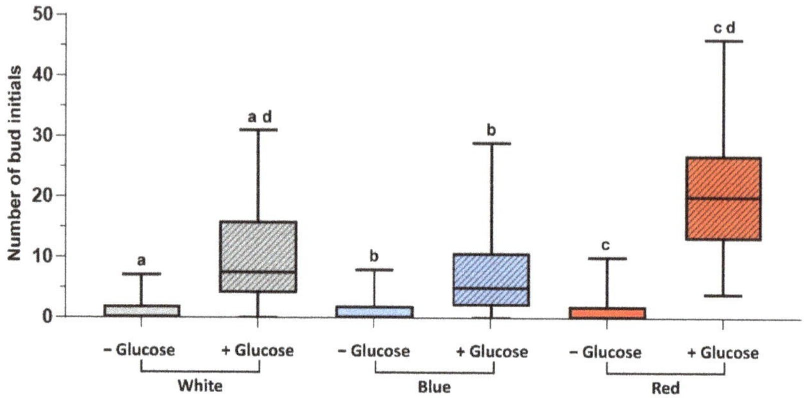

Figure 3. Box plots showing the comparison of bud initials formed in the presence and absence of glucose under different light conditions (in presence of 1 μM kinetin). In the absence of glucose very less number of bud initials or no bud initials are formed compared to their number in the presence of glucose. When glucose is present, maximum number of bud initials are formed in RL, which is significantly higher compared to the control (WL). Similar alphabetical letters indicate the significant difference between the bud initial numbers in the given condition ($p < 0.05$).

In all light conditions protonema growth was accelerated in the presence of glucose (Figure 1). However, under WL and BL, protonema showed predominant growth of chloronema irrespective of the presence or absence of glucose (Figure 2(i)a,b,d,e,g,h,j,k). On the other hand, protonema cultured under RL in the presence of glucose showed the emergence of caulonema filaments (Figure 2(i)i,l,(ii)i,l), but protonema cultured in the absence of glucose under RL did not show the emergence of caulonema filaments. Interestingly these protonemata (in the absence of glucose under RL) showed curling of newly-formed protonemal branches or nutations (Figure 2(i)c,f,(ii)c,f). The outline of these protonema colonies did not show protruding filaments due to curling of the protonemata (Figure 1c,f), which was observed in other conditions described earlier (Figure 1). Similar response has been described in *Funaria* under RL, but nutations were formed in presence of glucose, which is in contrast to *Physcomitrium*. These nutations later grew in a straight pattern in *Funaria* [52].

The curling of protonema filaments was also tested in WT cultures under BR light in medium with and without kinetin (1 μM) but lacking glucose. The response was compared with cultures under RL. It was observed that in BR light, protonema did not show nutations (Figure 4).

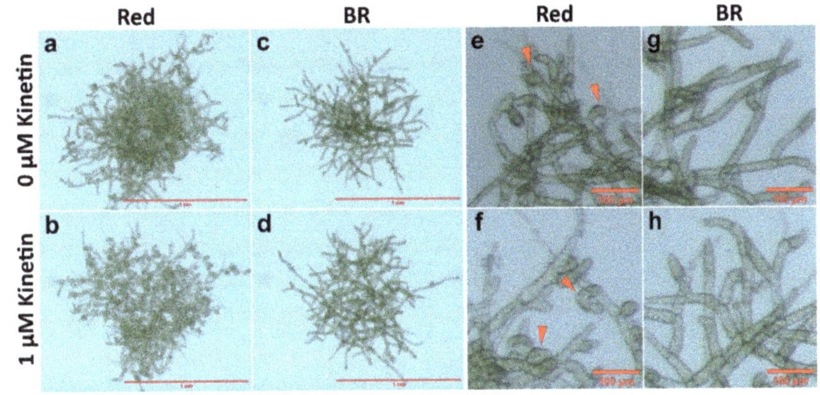

Figure 4. Comparison of WT protonema morphology under RL and BR light in the absence of glucose. **a–d** display complete protonema colony. **e–h** display magnified portions of protonema. Peripheral regions show nutations under RL (**a**,**b**,**e**,**f**). Nutations are not observed in BR light (**c**,**d**,**g**,**h**). Red triangles indicate the nutations (**e**,**f**). For **a–d**, scale bar = 1 mm. For **e–h**, scale bar = 100 µm.

3.2. *cry1a* Mutants Produce More Number of Gametophore Buds Than WT

Bud initial formation was compared in WT and *cry1a* protonema in medium supplemented with glucose and kinetin under WL, BL, RL, and BR light (Figure 5). Under WL, BL, and BR *cry1a* protonema produced a significantly greater number of bud initials than WT. On the other hand, there was no significant difference in the number of buds produced between WT and *cry1a* mutants under RL (Figure 6).

WT plants showed no significant difference in the number of bud initials produced under WL and BR. A similar trend was observed in *cry1a* protonema, which also showed no significant difference in the bud initial numbers between WL and BR. On the other hand, while WT plants exhibit no significant difference in bud initial numbers under WL and BL, *cry1a* plants produced significantly greater number of bud initials under WL compared to BL. Both WT and *cry1a* protonema produced significantly greater number of bud initials under RL than BR light. When the bud initial numbers were compared between BL and BR, WT plants produced a comparable number of bud initials, but *cry1a* plants produced significantly higher number of bud initials under BR (Figure 6). Overall comparison indicates that *cry1a* plants produce more number of gametophore buds compared to WT plants.

Figure 5. (**i**) WT and *cry1a* protonema colonies cultured in medium supplemented with glucose and kinetin (1µM) under different light conditions. In WT colonies bud initials are sparse (except RL) compared to *cry1a* plants, where these are visible (Scale bar = 1 mm). (**ii**) Peripheral regions of protonema colonies showing bud initials. Bud initials in WT plants are shown with red triangles. *cry1a* plants have more bud initials that WT plants. Scale bar = 500 µm (0.5 mm). (**iii**) Peripheral regions of protonema colonies (magnified) showing bud initials (Scale bar = 100 µm).

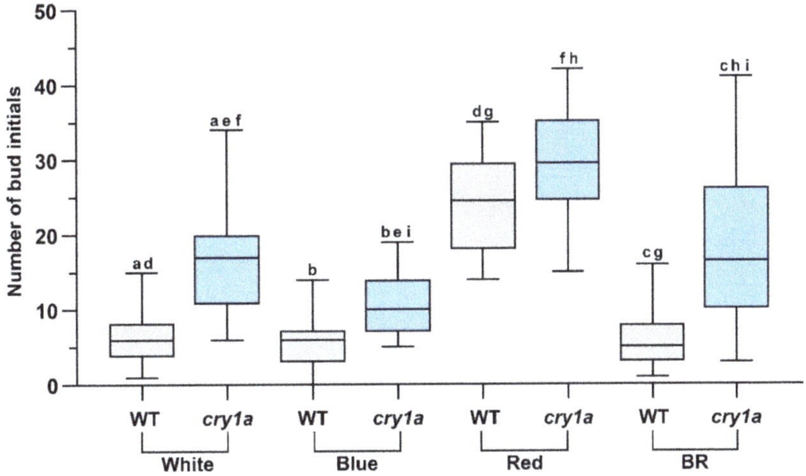

Figure 6. Comparison of number of bud initials in WT and *cry1a* protonema under different light conditions. Similar alphabetical letters indicate the significant difference between the bud initial numbers in the given condition ($p < 0.05$).

3.3. Gametophore Buds Are Not Formed under FR Light

Protonema cultured under FR light (in glucose supplemented media) in presence or absence of kinetin did not develop bud initials. However, they showed predominant growth of caulonema filaments under FR irrespective of the presence of kinetin (Figure 7c,d,g,h).

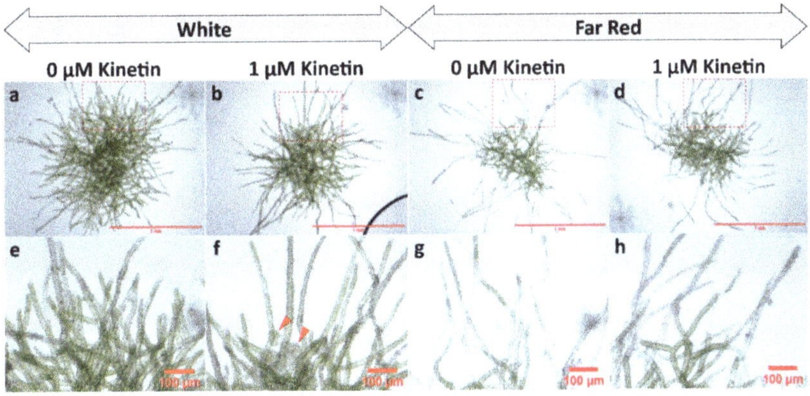

Figure 7. WT protonema colonies under WL and FR light in the presence of glucose. Upper panel shows the protonema colonies (**a**–**d**). Lower panel shows a portion of the peripheral region of the protonema colonies (magnified) (inset dashed red rectangle) displaying chloronema (**e**,**f**) or caulonema (**g**,**h**). Chloronemata are formed under WL (green filaments). Caulonemata are formed in FR light (less chlorophyll). Red triangles show the buds formed in WL in the presence of kinetin (**f**). For **a**–**d**, scale bar = 1 mm. For **e**–**h**, scale bar = 100 μm.

3.4. Phytohoromone Estimation

*t*Z content was estimated in WT and *cry1a* plants under WL. The difference between *t*Z content was not significant (Figure 8).

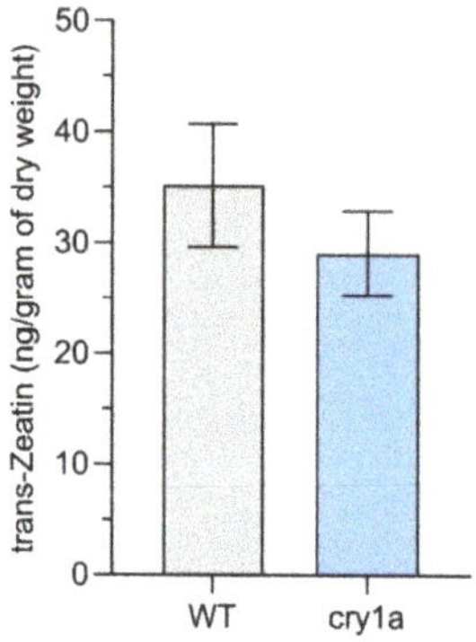

Figure 8. Comparison of *tZ* content in WT and *cry1a* plants. Comparison has been shown as mean ± SEM.

4. Discussion

4.1. Red Light and Glucose Enhance the Effect of Cytokinin

While light and carbon supply are known to influence auxin-mediated chloronema-caulonema transition in *Physcomitrium* [40,49], the influence of these factors on CK-mediated bud initial formation in *Physcomitrium* has not been clearly demonstrated.

In *Funaria* glucose induces CK-mediated gametophore bud initial formation [61]. In most of the previous studies involving CK response in mosses, carbon supply was provided either as glucose or sucrose. In the present study, bud initials were observed in WT protonema under WL, only when the culture medium was supplemented with both glucose and kinetin, but not in cultures lacking glucose and supplemented with kinetin alone. In the latter case either no bud initials were formed or rarely formed (Figure 2(ii)d–f). This indicates that optimum level of carbon source is required for CK response to initiate the bud initial formation.

Light quality also plays a role in the gametophore bud differentiation process. RL has been shown to have an enhancing effect on bud initial formation in different moss species. In this study, *Physcomitrium* also responded positively to RL as an inductive signal for bud initial formation, since the number of gametophore buds formed under RL was significantly higher compared to WL and BL (Figures 3 and 6).

WT protonema predominantly produced chloronema filaments in WL and BL, and protonema growth was accelerated in presence of glucose. However, there was no sign of caulonema or bud initial formation in WL and BL in the presence or absence of glucose. On the other hand, protonema under RL showed differential response to the presence of glucose. It is known that caulonema formation is enhanced under RL in *Physcomitrium* [40]. In the present study, caulonema filaments were produced under RL in medium supplemented with glucose, but not in absence of glucose (Figure 1c,f,i,l and Figure 2(i)c,f,i,l,(ii)c,f,i,l). Protonema cultured in presence of both glucose and kinetin under RL produced bud initials. Nutations were observed under RL in medium lacking glucose. Even kinetin did not stimulate gametophore bud formation in this condition. This observation has two implications for the impact of RL on protonema differentiation. First, RL is the primary signal for bud initial formation. While gametophore buds are formed to a differential extent in WL, BL, and RL, nutations are observed only under RL. Nutations later develop into buds. Therefore, RL plays a prominent role in gametophore bud formation in *Physcomitrium*. Second, RL also requires an optimum amount of energy in the form of carbon supply to promote the activity of CKs. In presence of glucose, RL induces the formation of caulonema (instead nutation) followed by gametophore buds. The energy requirement may be an upstream step/pre-requirement for bud initial formation via the formation of caulonema filaments in the natural environment (Figure 9i). The absence of nutations in WL and BL indicates that BL may suppress the effect of RL to induce the formation of nutations in *Physcomitrium*. To test this assumption WT protonema were cultured in BR light in medium lacking glucose, but in presence and absence of kinetin. Interestingly nutations were not observed in BR light in absence of glucose and the protonemata were phenotypically similar to those under WL (Figure 1a,d and Figure 4). This indicates that BL inhibits the formation of nutations in natural environment under WL, when energy level is not optimum (Figure 9ii). Since nutations are formed in presence of kinetin, it appears that this process is independent of CK action. However, it is not clear that the formation of nutations is regulated by any other phytohormone such as auxin. Gametophore bud initials are formed later when the cells synthesize the optimum carbon sources required for CK activity (Figure 9ii).

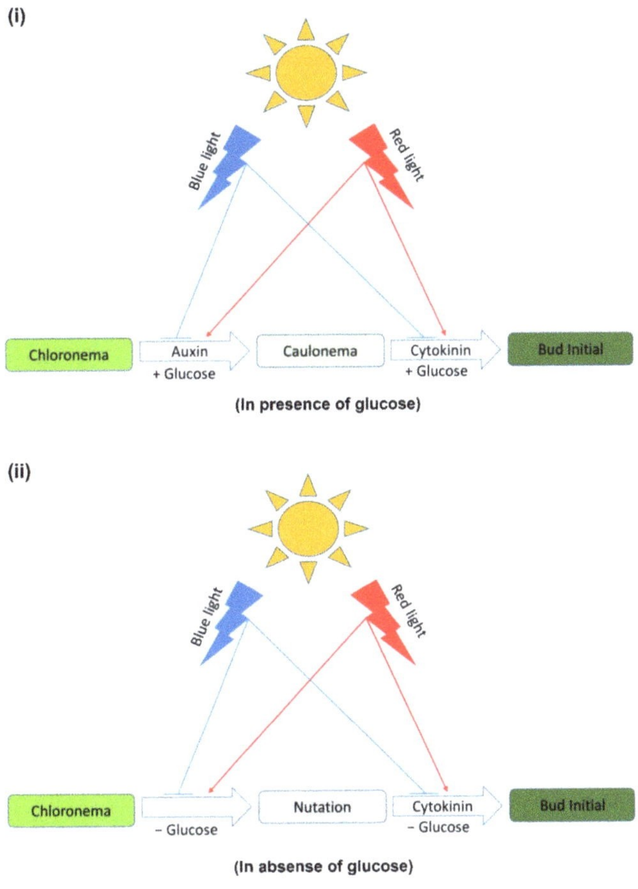

Figure 9. Graphical summary of the hypothesized mechanism of light and hormonal regulation of bud initial formation. (**i**) RL induces the formation of caulonema in presence of carbon sources such as glucose. Later bud initials are formed from caulonema branches. Glucose enhances the activity of auxin and CK. BL is inhibitory for the formation of both caulonema and bud initials (**ii**) RL induces formation of nutations in the absence of glucose. This process is inhibited by BL. Nutations later give rise to bud initials.

4.2. CRY1a *Suppresses Bud Initial Development*

Physcomitrium possesses two copies of CRYs [40]. The double disruptant *cry1a cry1b* mutant of *Physcomitrium* has been shown to produce more number of gametophores than either *cry1a* or *cry1b* mutant and WT plants produce an even lower number of gametophores under WL [40]. This implicates their role in the inhibition of bud initial formation [40].

However, in the same study WT, *cry1a* and *cry1b* were shown to produce a comparable number of gametophores with the *cry1a cry1b* strain under BL [40]. Therefore, it is not clear whether the biosynthesis or sensitivity of CKs is inhibited. In the current study WT protonema developed a comparable number of bud initials under WL and BL in response to exogenously supplied kinetin. We therefore also compared the budding response of WT and *cry1a* plants to exogenously supplied kinetin under different light conditions. No significant difference was observed in bud initial numbers between WT and *cry1a* protonema under RL (Figure 6). Since RL-induced bud initiation is phytochrome-mediated, CRY1a may not have a major impact under monochromatic RL.

The difference in the formation of bud initials became apparent, when WT and *cry1a* strains were cultured under WL, BL, and BR. Under WL, which is a mixture of different light wavelengths, both PHYs and CRYs become active. The greater number of buds in response to exogenously supplied kinetin in the *cry1a* mutant indicates that CRY1a may play a role in sensing the CKs in vivo. The observations under WL were further established from the budding response under BR light, where a similar pattern was observed. Since in natural sunlight spectrum plants respond to BL and RL, BR light presents a condition similar to WL. Under BL *cry1a* plants develop more number of bud initials than WT plants. These facts indicate that CRY1a suppresses bud initial formation. *cry1* protonema also produced more number of bud initials under WL and BR light compared to BL, but these differences were insignificant in WT plants. This indicates that CRY1a endogenously inhibits the CK response by inhibiting the RL effect.

Different types of CKs such as N^6-(Δ^2-isopentenyl)adenine (iP), *t*Z, *cis*-zeatin (*c*Z), dihydrozeatin (DHZ), and benzyl adenine (BA) are present in *Physcomitrium*. While cZ is the most abundant CK in *Physcomitrium*, iP, *t*Z, and BA display the most potent bud-inducing activity [6]. Since it is not known whether CRYs interfere with the sensitivity of the biosynthesis of CKs, *t*Z content was estimated in WT and *cry1a* plants cultured under WL. Both plants produce a comparable amount of *t*Z, but a differential amount of bud initials under WL. This indicates that CRYs may interfere with the CK-signaling. Biosynthesis and activity of other bioactive CKs might also be differentially regulated in *cry1a* plants of *Physcomitrium*. Since this study was conducted with a single CRY disruptant, there needs to be further investigation using double disruptant *cry1a cry1b* mutants to further establish the role of BL in CK-mediated bud formation.

4.3. FR Light Is Inhibitory for Bud Initial Development

No bud initials were developed in WT protonema in presence of kinetin. Protonema developed numerous caulonema under FR (Figure 7c,d,g,h), which may represent etiolated growth in protonema. It is known that caulonema formation is induced under RL [40], and it is required for bud initial formation. Caulonema formation also occurs under BL, but to a lesser extent compared to RL and it is almost absent under WL in standard culture conditions [40]. While overall growth is promoted under WL, BL, and RL, under FR fewer protonemata develop and more than 95% of the newly-formed protonemata are of caulonema type (unpublished data). This amount is much higher compared to RL as reported by Imaizumi et al. [40]. Calulonema formation has been shown to be induced under dark conditions and this shows red-FR reversibility where RL inhibits the dark-induced caulonema formation and FR light reverses this response [44]. While caulonema formation is followed by bud induction, bud initials were not developed in FR light even in the presence of glucose and kinetin (Figure 7). We observed that protonema cultured under FR light does not develop gametophores, but gametophores are produced (in lower numbers) when the protonema are pre-cultured in WL before their transfer to FR light (unpublished data). In WT plants of *Physcomitrium*, the number of gametophore bud initials decreases in a mixture of RL and FR light compared to RL, when the protonemata are cultured in kinetin-supplemented medium (unpublished data). These facts indicate that FR light is inhibitory for bud initial formation. Caulonema induction by RL and FR light perhaps represents two different developmental regulations mediated by auxin.

4.4. Light Quality and Carbon Supply May Influence the 2D–3D Transition in Physcomitrium

The evolution of 3D growth form is an important step in the colonization of land habit by plants. *Physcomitrium patens* is an excellent model to unravel the evolutionary innovations that facilitated the 2D–3D transition [62]. Since CKs induce the bud initial formation, the first 3D structures, they are important regulators of 2D–3D transition. The activity of CKs in plants are regulated by their synthesis, signaling, and degradation or inactivation [63]. ISOPENTENYLTRANSFERASEs (IPTs) are enzymes playing a major role in the rate limiting step of CK biosynthesis. *Physcomitrium* possesses seven copies of IPTs. *ipt1* knockout plants of *Physcomitrium* appear to have no defect in CK signaling, but show abundance of iP-type CK instead of *c*Z [64]. CK is perceived by CHASE domain-containing histidine kinases (CHKs). *Physcomitrium* encodes three CHKs namely PpCHK1, PpCHK2, and PpCHK3 [65]. *chk* mutants show a defect in gametophore development and differential budding in response to exogenous CK. Triple *chk* mutants show strong insensitivity to exogenous CK [65]. Cytokinin oxidases/dehydrogenases (CKXs) are enzymes catalyzing the degradation of CKs. In *Physcomitrium* the CKX gene family consists of six members [66]. Overexpression of *PpCKX1* results in delayed gametophore formation [66]. Since CK response is differentially regulated by light, the genes involved in CK biosynthesis, signaling, and degradation might also be differentially expressed under different light conditions.

In recent years numerous proteins have been identified which regulate the transition of 2D–3D growth downstream of CK signaling pathway in *Physcomitrium* [62]. These include CELLULOSE SYNTHASE 5 (CESA5), NO GAMETOPHORE 1 (PpNOG1), PpNOG2, DEFECTIVE KERNEL 1 (PpDEK1), AINTEGUMENTA, PLETHORA *and* BABY-BOOM (PpAPB), CLAVATA 3-like (CLE3) [62,67–72]. Some of these proteins are also regulated by auxin [62]. The mutants of these proteins show defect in gametophore bud formation.

We have shown that gametophore bud formation is regulated by light quality and energy supply. Therefore, it can be presumed that the components of CK response pathway as well as the proteins involved in 2D–3D transition might also be differentially regulated by these factors. Further studies are required to shed light on this aspect of CK response regulation.

5. Conclusions

Physcomitrium development has been shown to be regulated by the interaction of light and hormones. In addition, energy status also plays an important role in this differentiation process. BL and RL regulate the protonema differentiation process in an opposing manner, which is the growth and division in two dimensions. In this study we showed that light quality also influences the bud initial formation, which is the transition to 3D growth. Carbon supply appears to complement the effects of RL or PHYs. It is noteworthy that *Physcomitrium* possesses seven copies of PHYs. However, the role of individual PHYs in regulating the growth and development of *Physcomitrium* is not understood. We also showed that CRY1a may negatively regulate PHY action and thus CK response. Since *Physcomitrium* has another homolog of CRY1a, i.e., CRY1b, some of the responses which may be redundant, are difficult to interpret using the *cry1a* plants alone. Further studies involving higher order photoreceptor mutants are required to unravel the unknown aspects of light-regulation of the hormonal pathways in *Physcomitrium*.

Author Contributions: K.C.S.P. and D.P.B. conceptualized the idea. D.P.B. designed and performed the experiments, analyzed data, and wrote the manuscript. K.C.S.P. corrected the manuscript. All authors have read and agreed to the published version of the manuscript.

Funding: K.C.S.P. acknowledges funding from Department of Atomic Energy (DAE), Government of India and Council of Scientific and Industrial Research (CSIR), India (Grant No. 38(1495)/19/EMR-II) for funding. D.P.B. acknowledges Homi Bhabha National Institute (HBNI) for fellowship.

Institutional Review Board Statement: Not applicable.

Informed Consent Statement: Not applicable.

Data Availability Statement: Not applicable.

Acknowledgments: The authors are grateful to Mitsuyasu Hasebe (National Institute for Basic Biology, Japan) for providing the *cry1a* mutant. The authors thank Sony Kumari (NISER, Bhubaneswar) for suggesting corrections.

Conflicts of Interest: The authors declare no conflict of interest.

References

1. Rensing, S.A.; Goffinet, B.; Meyberg, R.; Wu, S.-Z.; Bezanilla, M. The moss *Physcomitrium* (*Physcomitrella*) *patens*: A model organism for non-seed plants. *Plant Cell* **2020**, *32*, 1361–1376. [CrossRef]
2. Beike, A.K.; von Stackelberg, M.; Schallenberg-Rüdinger, M.; Hanke, S.T.; Follo, M.; Quandt, D.; McDaniel, S.F.; Reski, R.; Tan, B.C.; Rensing, S.A. Molecular evidence for convergent evolution and allopolyploid speciation within the *Physcomitrium-Physcomitrella* species complex. *BMC Evol. Biol.* **2014**, *14*, 158. [CrossRef]
3. Medina, R.; Johnson, M.G.; Liu, Y.; Wickett, N.J.; Shaw, A.J.; Goffinet, B. Phylogenomic delineation of *Physcomitrium* (Bryophyta: Funariaceae) based on targeted sequencing of nuclear exons and their flanking regions rejects the retention of *Physcomitrella*, *Physcomitridium* and *Aphanorrhegma*. *J. Syst. Evol.* **2019**, *57*, 404–417. [CrossRef]
4. Takezawa, D.; Komatsu, K.; Sakata, Y. ABA in bryophytes: How a universal growth regulator in life became a plant hormone? *J. Plant Res.* **2011**, *124*, 437–453. [CrossRef]
5. Thelander, M.; Landberg, K.; Sundberg, E. Auxin-mediated developmental control in the moss *Physcomitrella patens*. *J. Exp. Bot.* **2018**, *69*, 277–290. [CrossRef]
6. von Schwartzenberg, K.; Núñez, M.F.; Blaschke, H.; Dobrev, P.I.; Novák, O.; Motyka, V.; Strnad, M. Cytokinins in the bryophyte *Physcomitrella patens*: Analyses of activity, distribution, and cytokinin oxidase/dehydrogenase overexpression reveal the role of extracellular cytokinins. *Plant Physiol.* **2007**, *145*, 786–800. [CrossRef]
7. Guillory, A.; Bonhomme, S. Phytohormone biosynthesis and signaling pathways of mosses. *Plant Mol. Biol.* **2021**, *107*, 245–277. [CrossRef]
8. Hayashi, K.; Horie, K.; Hiwatashi, Y.; Kawaide, H.; Yamaguchi, S.; Hanada, A.; Nakashima, T.; Nakajima, M.; Mander, L.N.; Yamane, H.; et al. Endogenous diterpenes derived from *ent*-kaurene, a common gibberellin precursor, regulate protonema differentiation of the moss *Physcomitrella patens*. *Plant Physiol.* **2010**, *153*, 1085–1097. [CrossRef]
9. Hayashi, K.; Kawaide, H.; Notomi, M.; Sakigi, Y.; Matsuo, A.; Nozaki, H. Identification and functional analysis of bifunctional *ent*-kaurene synthase from the moss *Physcomitrella patens*. *FEBS Lett.* **2006**, *580*, 6175–6181. [CrossRef]
10. Coudert, Y.; Palubicki, W.; Ljung, K.; Novak, O.; Leyser, O.; Harrison, C.J. Three ancient hormonal cues co-ordinate shoot branching in a moss. *Elife* **2015**, *4*, e06808. [CrossRef]
11. Yasumura, Y.; Pierik, R.; Fricker, M.D.; Voesenek, L.A.C.J.; Harberd, N.P. Studies of *Physcomitrella patens* reveal that ethylene-mediated submergence responses arose relatively early in land-plant evolution. *Plant J.* **2012**, *72*, 947–959. [CrossRef]
12. Johri, M.M.; Desai, S. Auxin regulation of caulonema formation in moss protonema. *Nat. New Biol.* **1973**, *245*, 223–224. [CrossRef]
13. Ashton, N.W.; Grimsley, N.H.; Cove, D.J. Analysis of gametophytic development in the moss, *Physcomitrella patens*, using auxin and cytokinin resistant mutants. *Planta* **1979**, *144*, 427–435. [CrossRef]
14. Bhatla, S.C.; Bopp, M. The hormonal regulation of protonema development in mosses III. Auxin-resistant mutants of the moss *Funaria hygrometrica* Hedw. *J. Plant Physiol.* **1985**, *120*, 233–243. [CrossRef]
15. Bhatla, S.C.; Kapoor, S.; Khurana, J.P. Involvement of calcium in auxin-induced cell differentiation in the protonema of the wild strain and auxin mutants of the moss *Funaria hygrometrica*. *J. Plant Physiol.* **1996**, *147*, 547–552. [CrossRef]
16. Johri, M.M. Caulonema differentiation in *Funaria protonema*. *Int. J. Dev. Biol.* **2020**, *64*, 21–28. [CrossRef]
17. Hoffmann, B.; Proust, H.; Belcram, K.; Labrune, C.; Boyer, F.D.; Rameau, C.; Bonhomme, S. Strigolactones inhibit caulonema elongation and cell division in the moss *Physcomitrella patens*. *PLoS ONE* **2014**, *9*, e99206. [CrossRef]
18. Mitra, G.C.; Allsopp, A. Effects of kinetin, gibberellic acid and certain auxins on the development of shoot buds on the protonema of *Pohlia nutans*. *Nature* **1959**, *183*, 974–975. [CrossRef]
19. Szweykowska, A. Kinetin-induced formation of gametophores in dark cultures of *Ceratodon purpureus*. *J. Exp. Bot.* **1963**, *14*, 137–141. [CrossRef]
20. Brandes, H.; Kende, H. Studies on cytokinin-controlled bud formation in moss protonemata. *Plant Physiol.* **1968**, *43*, 827–837. [CrossRef]
21. Szweykowska, A.; Guzowska, I.; Gallas, J. Studies on the activity of kinetin in cultures of *Funaria hygrometrica* II. Effect of kinetin on isolated protonema cells. *Acta Soc. Bot. Pol.* **1968**, *37*, 201–206. [CrossRef]
22. Chopra, R.N.; Bhatla, S.C. Regulation of gametangial formation in bryophytes. *Bot. Rev.* **1983**, *49*, 29–63. [CrossRef]
23. Chopra, R.N.; Dhingra-Babbar, S. Studies on bud induction in the moss *Trematodon brevicalyx* Dixon. *New Phytol.* **1984**, *97*, 613–620. [CrossRef]
24. Bopp, M.; Jacob, H.J. Cytokinin effect on branching and bud formation in *Funaria*. *Planta* **1986**, *169*, 462–464. [CrossRef]
25. Arif, M.A.; Hiss, M.; Tomek, M.; Busch, H.; Meyberg, R.; Tintelnot, S.; Reski, R.; Rensing, S.A.; Frank, W. ABA-induced vegetative diaspore formation in *Physcomitrella patens*. *Front. Plant Sci.* **2019**, *10*, 315. [CrossRef]

26. Yasumura, Y.; Pierik, R.; Kelly, S.; Sakuta, M.; Voesenek, L.A.C.J.; Harberd, N.P. An ancestral role for CONSTITUTIVE TRIPLE RESPONSE1 proteins in both ethylene and abscisic acid signaling. *Plant Physiol.* **2015**, *169*, 283–298. [CrossRef]
27. Biswal, D.P.; Panigrahi, K.C.S. Light- and hormone-mediated development in non-flowering plants: An overview. *Planta* **2020**, *253*, 1. [CrossRef]
28. de Wit, M.; Galvão, V.C.; Fankhauser, C. Light-mediated hormonal regulation of plant growth and development. *Annu. Rev. Plant Biol.* **2016**, *67*, 513–537. [CrossRef]
29. Galvão, V.C.; Fankhauser, C. Sensing the light environment in plants: Photoreceptors and early signaling steps. *Curr. Opin. Neurobiol.* **2015**, *34*, 46–53. [CrossRef] [PubMed]
30. Mathews, S. Phytochrome-mediated development in land plants: Red light sensing evolves to meet the challenges of changing light environments. *Mol. Ecol.* **2006**, *15*, 3483–3503. [CrossRef]
31. Han, Y.J.; Song, P.S.; Kim, J.I. Phytochrome-mediated photomorphogenesis in plants. *J. Plant Biol.* **2007**, *50*, 230–240. [CrossRef]
32. Wang, Q.; Lin, C. Mechanisms of cryptochrome-mediated photoresponses in plants. *Annu. Rev. Plant Biol.* **2020**, *71*, 103–129. [CrossRef] [PubMed]
33. Kinoshita, T.; Doi, M.; Suetsugu, N.; Kagawa, T.; Wada, M.; Shimazaki, K. phot1 and phot2 mediate blue light regulation of stomatal opening. *Nature* **2001**, *414*, 656–660. [CrossRef] [PubMed]
34. Rizzini, L.; Favory, J.J.; Cloix, C.; Faggionato, D.; O'Hara, A.; Kaiserli, E.; Baumeister, R.; Schäfer, E.; Nagy, F.; Jenkins, G.I.; et al. Perception of UV-B by the *Arabidopsis* UVR8 protein. *Science* **2011**, *332*, 103–106. [CrossRef] [PubMed]
35. Somers, D.E.; Schultz, T.F.; Milnamow, M.; Kay, S.A. *ZEITLUPE* encodes a novel clock-associated PAS protein from *Arabidopsis*. *Cell* **2000**, *101*, 319–329. [CrossRef]
36. Paik, I.; Huq, E. Plant photoreceptors: Multi-functional sensory proteins and their signaling networks. *Semin. Cell Dev. Biol.* **2019**, *92*, 114–121. [CrossRef]
37. van Gelderen, K.; Kang, C.; Pierik, R. Light signaling, root development, and plasticity. *Plant Physiol.* **2018**, *176*, 1049–1060. [CrossRef]
38. Kong, S.G.; Okajima, K. Diverse photoreceptors and light responses in plants. *J. Plant Res.* **2016**, *129*, 111–114. [CrossRef]
39. Li, F.-W.; Melkonian, M.; Rothfels, C.J.; Villarreal, J.C.; Stevenson, D.W.; Graham, S.W.; Wong, G.K.S.; Pryer, K.M.; Mathews, S. Phytochrome diversity in green plants and the origin of canonical plant phytochromes. *Nat. Commun.* **2015**, *6*, 7852. [CrossRef]
40. Imaizumi, T.; Kadota, A.; Hasebe, M.; Wada, M. Cryptochrome light signals control development to suppress auxin sensitivity in the moss *Physcomitrella patens*. *Plant Cell* **2002**, *14*, 373–386. [CrossRef]
41. Kasahara, M.; Kagawa, T.; Sato, Y.; Kiyosue, T.; Wada, M. Phototropins mediate blue and red light-induced chloroplast movements in *Physcomitrella patens*. *Plant Physiol.* **2004**, *135*, 1388–1397. [CrossRef] [PubMed]
42. Soriano, G.; Cloix, C.; Heilmann, M.; Núñez-Olivera, E.; Martínez-Abaigar, J.; Jenkins, G.I. Evolutionary conservation of structure and function of the UVR8 photoreceptor from the liverwort *Marchantia polymorpha* and the moss *Physcomitrella patens*. *New Phytol.* **2018**, *217*, 151–162. [CrossRef] [PubMed]
43. Linde, A.-M.; Eklund, D.M.; Kubota, A.; Pederson, E.R.A.; Holm, K.; Gyllenstrand, N.; Nishihama, R.; Cronberg, N.; Muranaka, T.; Oyama, T.; et al. Early evolution of the land plant circadian clock. *New Phytol.* **2017**, *216*, 576–590. [CrossRef] [PubMed]
44. Cove, D.J.; Schild, A.; Ashton, N.W.; Hartmann, E. Genetic and physiological studies of the effect of light on the development of the moss, *Physcomitrella patens*. *Photochem. Photobiol.* **1978**, *27*, 249–254. [CrossRef]
45. Possart, A.; Fleck, C.; Hiltbrunner, A. Shedding (far-red) light on phytochrome mechanisms and responses in land plants. *Plant Sci.* **2014**, *217–218*, 36–46. [CrossRef]
46. Mittmann, F.; Brücker, G.; Zeidler, M.; Repp, A.; Abts, T.; Hartmann, E.; Hughes, J. Targeted knockout in *Physcomitrella* reveals direct actions of phytochrome in the cytoplasm. *Proc. Natl. Acad. Sci. USA* **2004**, *101*, 13939–13944. [CrossRef]
47. Bao, L.; Yamamoto, K.T.; Fujita, T. Phototropism in gametophytic shoots of the moss *Physcomitrella patens*. *Plant Signal. Behav.* **2015**, *10*, e1010900. [CrossRef]
48. Lymperopoulos, P.; Msanne, J.; Rabara, R. Phytochrome and phytohormones: Working in tandem for plant growth and development. *Front. Plant Sci.* **2018**, *9*, 1037. [CrossRef]
49. Thelander, M.; Olsson, T.; Ronne, H. Effect of the energy supply on filamentous growth and development in *Physcomitrella patens*. *J. Exp. Bot.* **2005**, *56*, 653–662. [CrossRef]
50. Mitra, G.C.; Misra, L.P.; Prabha, C. Interaction of red and blue light on the development of the protonema and bud formation in *Pohlia nutans*. *Planta* **1965**, *65*, 42–48. [CrossRef]
51. Mitra, G.C.; Misra, L.P.; Kaul, K.N. Effects of coconut milk and kinetin on the development of the protonema and bud formation in *Pohlia nutans*. *Nature* **1962**, *195*, 1219–1220. [CrossRef]
52. Simon, P.E.; Naef, J.B. Light dependency of the cytokinin-induced bud initiation in protonemata of the moss *Funaria hygrometrica*. *Physiol. Plant.* **1981**, *53*, 13–18. [CrossRef]
53. Chopra, R.N.; Gupta, U. Dark-induction of buds in *Funaria hygrometrica* Hedw. *Bryologist* **1967**, *70*, 102–104. [CrossRef]
54. Nebel, B.J.; Naylor, A.W. Light, temperature and carbohydrate requirements for shoot-bud initiation from protonemata in the moss *Physcomitrium turbinatum*. *Am. J. Bot.* **1968**, *55*, 38–44. [CrossRef]
55. Zavala, A.M.; Pérez, N.M.; Becerra, A.A.; López, M.A.V. Glucose inhibits the shoot bud formation in the moss *Bryum billarderi*. *Cent. Eur. J. Biol.* **2012**, *7*, 648–654. [CrossRef]

56. Sabovljevic, A.; Sabovljevic, M.; Grubisic, D.; Konjevic, R. The effect of sugars on development of two moss species (*Bryum argentum* and *Atrichum undulatum*) during in vitro culture. *Belg. J. Bot.* **2005**, *138*, 79–84.
57. Bopp, M.; Brandes, H. Versuche zur analyse der protonema entwicklung der laubmoose: II. Über den zusammenhang zwischen protonema differenzierung und kinetin wirkung bei der bildung von moos knospen. *Planta* **1964**, *62*, 116–136. [CrossRef]
58. Reski, R.; Abel, W.O. Induction of budding on chloronemata and caulonemata of the moss, *Physcomitrella patens*, using isopentenyladenine. *Planta* **1985**, *165*, 354–358. [CrossRef]
59. Ashton, N.W.; Cove, D.J. The isolation and preliminary characterisation of auxotrophic and analogue resistant mutants of the moss, *Physcomitrella patens*. *Mol. Gen. Genet.* **1977**, *154*, 87–95. [CrossRef]
60. Šimura, J.; Antoniadi, I.; Široká, J.; Tarkowská, D.; Strnad, M.; Ljung, K.; Novák, O. Plant hormonomics: Multiple phytohormone profiling by targeted metabolomics. *Plant Physiol.* **2018**, *177*, 476–489. [CrossRef]
61. Szweykowska, A.; Szweykowska, A.; Ratajczak, W.; Schneider, J. Studies on the activity of kinetin in cultures of *Funaria hygrometrica* I. Bud-inducing activity of kinetin and protein synthesis. *Acta Soc. Bot. Pol.* **1968**, *37*, 145–155. [CrossRef]
62. Moody, L.A. The 2D to 3D growth transition in the moss *Physcomitrella patens*. *Curr. Opin. Plant Biol.* **2019**, *47*, 88–95. [CrossRef] [PubMed]
63. Kieber, J.J.; Schaller, G.E. Cytokinin signaling in plant development. *Development* **2018**, *145*, dev149344. [CrossRef] [PubMed]
64. Lindner, A.C.; Lang, D.; Seifert, M.; Podlešáková, K.; Novák, O.; Strnad, M.; Reski, R.; von Schwartzenberg, K. Isopentenyltransferase-1 (IPT1) knockout in *Physcomitrella* together with phylogenetic analyses of IPTs provide insights into evolution of plant cytokinin biosynthesis. *J. Exp. Bot.* **2014**, *65*, 2533–2543. [CrossRef]
65. von Schwartzenberg, K.; Lindner, A.C.; Gruhn, N.; Šimura, J.; Novák, O.; Strnad, M.; Gonneau, M.; Nogué, F.; Heyl, A. CHASE domain-containing receptors play an essential role in the cytokinin response of the moss *Physcomitrella patens*. *J. Exp. Bot.* **2016**, *67*, 667–679. [CrossRef]
66. Hyoung, S.; Cho, S.H.; Chung, J.H.; So, W.M.; Cui, M.H.; Shin, J.S. Cytokinin oxidase PpCKX1 plays regulatory roles in development and enhances dehydration and salt tolerance in *Physcomitrella patens*. *Plant Cell Rep.* **2020**, *39*, 419–430. [CrossRef]
67. Moody, L.A.; Kelly, S.; Rabbinowitsch, E.; Langdale, J.A. Genetic regulation of the 2D to 3D growth transition in the moss *Physcomitrella patens*. *Curr. Biol.* **2018**, *28*, 473–478.e5. [CrossRef]
68. Demko, V.; Perroud, P.F.; Johansen, W.; Delwiche, C.F.; Cooper, E.D.; Remme, P.; Ako, A.E.; Kugler, K.G.; Mayer, K.F.X.; Quatrano, R.; et al. Genetic analysis of DEFECTIVE KERNEL1 loop function in three-dimensional body patterning in *Physcomitrella patens*. *Plant Physiol.* **2014**, *166*, 903–919. [CrossRef]
69. Whitewoods, C.D.; Cammarata, J.; Nemec Venza, Z.; Sang, S.; Crook, A.D.; Aoyama, T.; Wang, X.Y.; Waller, M.; Kamisugi, Y.; Cuming, A.C.; et al. CLAVATA was a genetic novelty for the morphological innovation of 3D Growth in land plants. *Curr. Biol.* **2018**, *28*, 2365–2376.e5. [CrossRef]
70. Moody, L.A.; Kelly, S.; Clayton, R.; Weeks, Z.; Emms, D.M.; Langdale, J.A. *NO GAMETOPHORES 2* is a novel regulator of the 2D to 3D growth transition in the moss *Physcomitrella patens*. *Curr. Biol.* **2021**, *31*, 555–563.e4. [CrossRef]
71. Goss, C.A.; Brockmann, D.J.; Bushoven, J.T.; Roberts, A.W. A CELLULOSE SYNTHASE (CESA) gene essential for gametophore morphogenesis in the moss *Physcomitrella patens*. *Planta* **2012**, *235*, 1355–1367. [CrossRef] [PubMed]
72. Aoyama, T.; Hiwatashi, Y.; Shigyo, M.; Kofuji, R.; Kubo, M.; Ito, M.; Hasebe, M. AP2-type transcription factors determine stem cell identity in the moss *Physcomitrella patens*. *Development* **2012**, *139*, 3120–3129. [CrossRef] [PubMed]

plants

MDPI

Article

Phenolic Compounds Content Evaluation of Lettuce Grown under Short-Term Preharvest Daytime or Nighttime Supplemental LEDs

Aušra Brazaitytė [1,*,†], Viktorija Vaštakaitė-Kairienė [1,†], Rūta Sutulienė [1], Neringa Rasiukevičiūtė [1], Akvilė Viršilė [1], Jurga Miliauskienė [1], Kristina Laužikė [1], Alma Valiuškaitė [1], Lina Dėnė [1], Simona Chrapačienė [1], Asta Kupčinskienė [1] and Giedrė Samuolienė [1,2]

1 Lithuanian Research Centre for Agriculture and Forestry, Institute of Horticulture, Kaunas Str. 30, LT-54333 Babtai, Lithuania; viktorija.vastakaite-kairiene@lammc.lt (V.V.-K.); ruta.sutuliene@lammc.lt (R.S.); neringa.rasiukeviciute@lammc.lt (N.R.); akvile.virsile@lammc.lt (A.V.); jurga.miliauskiene@lammc.lt (J.M.); kristina.lauzike@lammc.lt (K.L.); alma.valiuskaite@lammc.lt (A.V.); lina.dene@lammc.lt (L.D.); simona.chrapaciene@lammc.lt (S.C.); asta.kupcinskiene@lammc.lt (A.K.); giedre.samuoliene@lammc.lt (G.S.)
2 Department of Agroecosystems and Soil Sciences, Agriculture Academy, Vytautas Magnus University, Donelaicio Str. 58, 44248 Kaunas, Lithuania
* Correspondence: ausra.brazaityte@lammc.lt
† These authors contributed equally to this work.

Abstract: The study aimed to determine the changes in phenolic compounds content in lettuce (*Lactuca sativa* L. cv. Little Gem) depending on the preharvest short-term daytime or nighttime supplemental light-emitting diodes (LEDs) to high-pressure sodium lamps (HPS) lighting in a greenhouse during autumn and spring cultivation. Plants were grown in a greenhouse under HPS supplemented with 400 nm, 455 nm, 530 nm, 455 + 530 nm or 660 nm LEDs light for 4 h five days before harvest. Two experiments (EXP) were performed: EXP1—HPS, and LEDs treatment during daytime 6 PM–10 PM, and EXP2—LEDs treatment at nighttime during 10 AM–2 PM. LEDs' photosynthetic photon flux density (PPFD) was 50 and HPS—90 ± 10 μmol m^{-2} s^{-1}. The most pronounced positive effect on total phenolic compounds revealed supplemental 400 and 455 + 530 nm LEDs lighting, except its application during the daytime at spring cultivation, when all supplemental LEDs light had no impact on phenolics content variation. Supplemental 400 nm LEDs applied in the daytime increased chlorogenic acid during spring and chicoric acid during autumn cultivation. 400 nm LEDs used in nighttime enhanced chlorogenic acid accumulation and rutin during autumn. Chicoric and chlorogenic acid significantly increased under supplemental 455 + 530 nm LEDs applied at daytime in autumn and used at nighttime—in spring. Supplemental LEDs application in the nighttime resulted in higher phenolic compounds content during spring cultivation and the daytime during autumn cultivation.

Keywords: *Lactuca sativa* L.; light-emitting diodes; phenolic acids; flavonoids

Citation: Brazaitytė, A.; Vaštakaitė-Kairienė, V.; Sutulienė, R.; Rasiukevičiūtė, N.; Viršilė, A.; Miliauskienė, J.; Laužikė, K.; Valiuškaitė, A.; Dėnė, L.; Chrapačienė, S.; et al. Phenolic Compounds Content Evaluation of Lettuce Grown under Short-Term Preharvest Daytime or Nighttime Supplemental LEDs. *Plants* **2022**, *11*, 1123. https://doi.org/10.3390/plants11091123

Academic Editors: Hail Rihan and Mick Fuller

Received: 15 March 2022
Accepted: 20 April 2022
Published: 21 April 2022

1. Introduction

Nowadays, a growing interest in green eating is observed, affecting the increasing consumption of vegetables, including leafy vegetables. Therefore, not only their yield but also their nutritional quality becomes essential. Leafy vegetables are rich in bioactive secondary metabolites, which have health-beneficial properties for humans. Secondary metabolites participate in protecting plants against abiotic and biotic stresses and are essential for human nutrition, promoting the colour, taste, or aroma of plant products [1–3]. One of the leading secondary metabolites in plants are phenolic compounds. Such compounds, having antioxidant and anti-inflammatory properties, can help against the development of cancers, cardiovascular and neurodegenerative diseases, diabetes, obesity, etc. [3–5].

Because many leafy vegetables are grown in controlled environment agriculture (CEA), the synthesis and accumulation of secondary metabolites depend on the regulation of microclimate, remarkably light [1,3]. Plants react to light through multiple photoreceptors, which respond to a broad light spectrum, from ultraviolet B (UV-B) to far-red wavelengths and stimulate the biochemical pathways of such metabolites by regulating the expression of specific genes [2,3,6]. Moreover, in some cases, light could act as eustress (positive stress), stimulating the production of various phytochemicals, thus improving the nutritional value of leafy vegetables [1,3,7]. Nowadays, the application of the ecologically friendly technology of light-emitting diode (LED) lighting in CEA with its capability to select light wavelengths, change intensity, and reduce energy costs has many advantages compared to other conventional light sources. The ability to tailor the spectral composition according to plant or photoreceptor characteristics can affect leafy vegetables' primary and secondary metabolic responses [3,8–13]. Moreover, monochromatic LEDs or their combination can be used to supplement the spectrum of high-pressure sodium (HPS) or fluorescent (FL) lamps, which are still widely applied in CEA, throughout the cultivation of vegetables or as short-term pre-harvest exposure [7,14–20]. Few studies concerning LEDs' short-term pre-harvest exposure differ in the used spectrum, intensity, exposure time, and duration, and their effects on various phytochemical changes in leafy vegetables [7,15–18,20]. For example, short-term five days UV-A LEDs exposure increased antioxidant phenolic compounds in kale [20]. Short-term five days blue LED pre-harvest treatment significantly increased some carotenoids and glucosinolates in sprouting broccoli microgreens [21], and ten days exposure resulted in higher vitamin C, soluble protein, free amino acids, and chlorophyll in Chinese kale at harvest [22]. Continuous 48 h red-blue light-emitting diodes illumination depending on their ratio or intensity decreased nitrate and increased soluble sugars and vitamin C content [17,18]. Short-term pre-harvest red LED lighting exposure was shown as an efficient tool to reduce nitrate contents in various leafy vegetables [7,15] and produced baby leaf lettuce and *Brassicaceae* microgreens rich in total phenolics, tocopherols, sugars, and antioxidant capacity [16,23]. However, there is still a lack of information on how LEDs short-term pre-harvest exposure affects changes in the different phenolic compounds content in leafy vegetables. According to literature data, generally, different monochromatic light can affect the stimulation of secondary metabolites, especially phenolic compounds [2,3,24].

Furthermore, literature data showed that the growing season has also affected phytochemicals content in leafy vegetables [25]. Although vegetables are mainly grown in the CEA during the autumn-spring season, where supplemental lighting is used, limited data from the literature indicate the effect of seasonality. For example, after short-term red LED exposure, total phenolics content and DPPH free radical scavenging capacity within baby lettuces significantly increased during "dark" months, in November and January, but showed a different response in March [16]. Bioactive compounds increase in green baby leaf lettuces cultivar was observed in November and in red leaf cultivar in January under supplemental blue and green LEDs to HPS lighting [14]. The effect of seasonality on photosynthetic indices, growth, and phenolic compounds was determined in lamb's lettuce under various LEDs light combinations [26,27].

Lettuce (*Lactuca sativa* L.) is one of the highly valuable leafy vegetables cultivated in CEA, mainly for its fresh leaves. Lettuce varies in colors, sizes, shapes, and is mostly used in salad mixes. It is low in calories, fat, Na, a good source of fiber, minerals, various vitamins and bioactive compounds such as folate, vitamin E, vitamin C, β-carotene, and phenolic compounds. Phenolic acids, especially caffeic acid, chlorogenic acid, and their derivatives, and flavonoids such as quercetin and kaempferol derivatives, anthocyanins, and flavone luteolin are reported as the main phenolic compounds in lettuce [28]. Furthermore, literature data showed that phenolic compounds content in lettuce could be enhanced by manipulating various agricultural practices, including light through the application of LEDs [12,16,19,28]. For example, anthocyanin, flavonoids, and chlorogenic acids in red leaf lettuce significantly increased under a higher percentage of blue light [12]. In the same type of lettuce, the predawn application of blue light showed the highest content of phenolic acids and flavonoids in comparison with green leaf lettuce [19]. Supplemental red-LEDs before harvesting resulted in an increase in the total phenolics of baby leaf lettuce [16]. Therefore, we hypothesised that even short-term exposure of LEDs light as supplemental to HPS lighting would positively affect the content of phenolic compounds in lettuce depending on the time of day or season. Thus, our study aimed to determine the changes of phenolic compounds content in lettuce depending on the short-term daytime or nighttime preharvest supplemental LEDs to HPS lighting in a greenhouse during autumn and spring cultivation.

2. Results

2.1. Effect of Short-Term Daytime Supplemental LEDs to HPS Lighting on Phenolic Compounds Content in Lettuce Cultivated in A Greenhouse during Different Seasons

The results that short-term daytime preharvest supplemental LEDs to HPS lighting in a greenhouse during spring cultivation did not affect total content of phenolic compounds in lettuce (Table 1). There was a trend, that different lighting had different effects on the content of individual phenolic compounds in lettuce. Although daytime supplemental LEDs light during spring cultivation of lettuce reduced or did not affect the content of many phenolic compounds, chlorogenic acid was significantly increased under supplemental 400 nm LEDs, and rosmarinic acid—under 530 nm LEDs. Also, it was determined the positive effect of supplemental 455 nm LEDs on chlorogenic and rosmarinic acid content in lettuce. Meanwhile, compared to HPS lighting, supplemental 660 nm LEDs resulted in the lowest gallic acid and apigenin, myricetin, and rutin content, 530 nm—gallic acid and apigenin, epicatechin, quercetin, rutin, 455 + 530 nm—protocatechuic and rosmarinic acids, apigenin and myricetin, 455 nm—protocatechuic, kaempferol, 400 nm—rutin. All supplemental LEDs decreased caffeic and p-coumaric acids content.

During cultivation in autumn, the more evident effect of different lighting on the phenolic compounds in lettuce was noticed. Total phenol compounds content as well as phenolic acids such as caffeic, chicoric, chlorogenic, rosmarinic, and flavonol quercetin content were higher under supplemental 455 + 530 nm LEDs than under HPS lighting alone. Supplemental 660 nm LEDs significantly increased caffeic and o-coumaric acids, epicatechin, quercetin and rutin. Positive effects on the increase of chicoric and o-coumaric acids, epicatechin and quercetin, were determined under supplemental 400 nm and rosmarinic acid under 455 nm LEDs light. Although supplemental 530 nm LEDs light increased p-coumaric and rosmarinic acid content, the lowest content of caffeic, chlorogenic, protocatechuic acids and quercetin, and rutin was established. Different lighting had no significant effect on gallic acid, kaempferol, and myricetin content in lettuce.

The incidence of significant light and season interaction (LxS) indicates differential response to short-term daytime supplemental LEDs light treatments at spring and autumn examined with respect to phenolic compounds content (Table 1). However, the relative contribution of the main effects to the variance of phenolic compounds indicates that variation is introduced principally by season (S) (Table S2) and much less by light treatment (L) (Table S1). Season had no effect only on caffeic acid content. Higher levels of chicoric, chlorogenic and o-coumaric acids and epicatechin, quercetin, rutin and total phenolics were found during autumn cultivation (Table S2). Meanwhile, the effect of light treatment (L) was more pronounced on caffeic acid and epicatechin, rutin content, which were higher under the 660 nm supplemental LEDs treatments and quercetin content under 400 nm in comparison with other light treatments (Table S1).

Table 1. Effect of short-term daytime supplemental LEDs to HPS lighting on phenolic compounds content in lettuce cultivated in a greenhouse during different seasons.

Phenolic Compounds	Lighting												Source of Variance		
	HPS		400 nm		455 nm		455 + 530 nm		530 nm		660 nm		L	S	L × S
	Spring	Autumn	Spring	Autumn	Spring	Autumn	Spring	Autumn	Spring	Autumn	Spring	Autumn			
Caffeic a.	0.068 b	0.015 gh	0.033 ef	0.040 def	0.047 de	0.05 cd	0.029 efg	0.066 bc	0.025 fg	0.007 h	0.042 def	0.151 a	*		*
Chicoric a.	2.57 cd	3.89 bc	2.43 d	5.07 ab	2.24 d	3.83 bc	2.57 cd	5.51 a	2.26 d	3.87 bc	1.97 d	2.73 cd		*	*
Chlorogenic a.	0.81 de	1.04 bc	1.07 bc	0.98 cd	0.98 cd	1.16 bc	0.74 e	1.46 a	0.77 e	0.72 e	0.75 e	1.22 b		*	*
Gallic a.	0.049 a	0.029 bc	0.034 abc	0.020 bc	0.037 ab	0.019 bc	0.031 abc	0.022 bc	0.028 bc	0.018 c	0.026 bc	0.020 bc		*	*
o-coumaric a.	0.038 c	0.141 b	0.034 c	0.230 a	0.042 c	0.057 c	0.056 c	0.038 c	0.044 c	0.043 c	0.039 c	0.227 a		*	*
p-coumaric a	0.082 a	0.008 f	0.052 b	0.016 ef	0.039 c	0.016 ef	0.030 cd	0.010 ef	0.030 cd	0.033 c	0.021 de	0.016 ef		*	*
Protocatechuic a.	0.177 a	0.071 c	0.138 b	0.023 d	0.133 b	0.022 d	0.073 c	0.023 d	0.160 ab	0.020 d	0.142 ab	0.093 c		*	*
Rosmarinic a.	0.641 bc	0.049 f	0.596 c	0.076 f	0.732 a	0.520 cd	0.563 c	0.424 d	0.797 a	0.235 e	0.601 c	0.033 f		*	*
Apigenin	0.844 a	0.333 e	0.637 bc	0.082 f	0.725 ab	0.358 e	0.458 de	0.044 f	0.715 b	0.381 e	0.521 cd	0.089 f		*	*
Epicatechin	0.165 cd	0.280 b	0.080 de	0.432 a	0.069 de	0.137 cde	0.080 de	0.199 bc	0.047 e	0.187 bc	0.123 cde	0.515 a	*	*	*
Kaempferol	0.034 ab	0.020 cd	0.038 a	0.013 d	0.015 cd	0.026 bc	0.026 bc	0.020 cd	0.035 ab	0.024 bcd	0.021 cd	0.022 cd		*	*
Myricetin	0.118 a	0.051 c	0.089 b	0.035 c	0.107 ab	0.033 c	0.053 c	0.038 c	0.099 ab	0.037 c	0.052 c	0.040 c		*	*
Quercetin	0.043 d	0.022 de	0.036 de	0.180 a	0.026 de	0.015 e	0.036 de	0.179 a	0.071 c	0.014 e	0.028 de	0.139 b	*	*	*
Rutin	0.026 b	0.042 b	0.013 c	0.061 b	0.026 b	0.046 b	0.027 b	0.049 b	0.016 c	0.019 c	0.012 c	0.361 a	*	*	*
Total	5.67 cde	5.99 cd	5.28 cde	7.25 ab	5.21 cde	6.29 bc	4.78 de	8.08 a	5.10 cde	5.61 cde	4.34 e	5.65 cde		*	*

L—lighting; S—seasons; a.—acid. Individual phenolic compound content is presented as mg g^{-1} in dry plant matter. Means with different letters (based on heatmap values with the same letters on a separate line are marked with the same color) are significantly different at the $p < 0.05$ level by Tukey's honestly significant difference test (*).

2.2. Effect of Short-Term Nighttime Supplemental LEDs to HPS Lighting on Phenolic Compounds Content in Lettuce Cultivated in A Greenhouse during Different Seasons

According to data obtained from experiments, when short-term supplemental LEDs light were applied at nighttime, the most positive effect for total phenolic compounds increase was found under 455 + 530 nm LEDs during both cultivation periods and under 400 nm LEDs during autumn cultivation (Table 2). Generally, higher total phenolic compounds content in lettuce was determined under spring cultivation, contrary to what supplemental LEDs were applied in the daytime.

During spring cultivation, the supplemental 455 + 530 nm LEDs at nighttime increased the content of phenolic acids such as chicoric, chlorogenic, rosmarinic, and flavonol apigenin, which were the main part of the total phenolic compounds. Supplemental 455 nm LEDs light increased caffeic acid and 660 nm—kaempferol content. Different lighting had no significant effect on p-coumaric acid and rutin content. Gallic, o-coumaric, protocatechuic acids, and epicatechin decreased under all supplemental LEDs light. Supplemental 530 nm LEDs light resulted in the significantly lowest content of chicoric, chlorogenic, o-coumaric, and protocatechuic acids, 455 + 530 nm—caffeic acid, 455 nm—gallic acid and quercetin, 400 nm—epicatechin compared to HPS lighting.

Although nighttime supplemental LEDs light during autumn cultivation of lettuce did not affect many phenolic compounds content, chlorogenic, gallic, and o-coumaric acids, myricetin, and rutin increased under supplemental 400 nm LEDs as well as rutin under 455 + 530 nm LEDs. The positive effect of supplemental 530 nm LEDs was determined on the increase of caffeic acid. Meanwhile, supplemental 455 nm LEDs light significantly decreased chlorogenic acid and rutin content.

The same as at daytime application, significant light and season interaction (L $\times$ S) were determined on phenolic compounds content (Table 2). Season had no effect on o-coumaric acid and myricetin content. Higher content of total phenolics and practically all individual phenolic compounds, except o-coumaric acid and rutin were found during spring cultivation (Table S4). Meanwhile, the effect of light treatment (L) was more obvious on chlorogenic, o-coumaric acids, epicatechin and total phenolics content (Table S4).

Table 2. Effect of short-term nighttime supplemental LEDs to HPS lighting on phenolic compounds content in lettuce cultivated in a greenhouse during different seasons.

Phenolic Compounds	Lighting												Source of Variance		
	HPS		400 nm		455 nm		455 + 530 nm		530 nm		660 nm		L	S	L × S
	Spring	Autumn	Spring	Autumn	Spring	Autumn	Spring	Autumn	Spring	Autumn	Spring	Autumn			
Caffeic a.	0.222 b	0.036 ef	0.217 b	0.043 e	0.269 a	0.018 ef	0.116 d	0.039 e	0.181 c	0.098 d	0.180 c	0.009 f		*	*
Chicoric a.	4.47 b	0.28 e	2.40 d	0.40 e	3.65 c	0.08 e	5.42 a	0.37 e	1.78 d	0.29 e	1.89 d	0.13 e		*	*
Chlorogenic a.	2.59 b	1.11 de	1.39 d	2.15 c	2.04 c	0.33 f	5.08 a	1.44 d	1.25 d	1.12 de	2.01 c	0.80 e	*	*	*
Gallic a.	0.059 a	0.012 f	0.030 cd	0.036 bc	0.026 cde	0.016 def	0.049 ab	0.013 ef	0.044 b	0.009 f	0.027 cd	0.017 def		*	*
o-coumaric a.	0.303 ab	0.190 bc	0.043 c	0.382 a	0.070 bc	0.035 c	0.087 bc	0.047 c	0.035 c	0.115 bc	0.036 c	0.067 bc	*		*
p-coumaric a	0.021 ab	0.005 ab	0.021 ab	0.027 ab	0.025 ab	0.005 ab	0.052 a	0.003 b	0.024 ab	0.003 b	0.038 ab	0.004 ab		*	*
Protocatechuic a.	0.210 a	0.012 d	0.122 bc	0.022 d	0.101 c	0.013 d	0.148 b	0.019 d	0.098 c	0.022 d	0.156 b	0.029 d		*	*
Rosmarinic a.	0.098 de	0.006 e	0.723 c	0.015 e	1.153 b	0.007 e	2.318 a	0.005 e	0.453 cd	0.006 e	0.771 c	0.006 e		*	*
Apigenin	0.390 b	0.010 e	0.428 b	0.051 e	0.255 cd	0.044 e	0.597 a	0.010 e	0.298 c	0.009 e	0.196 d	0.061 e		*	*
Epicatechin	0.493 a	0.095 bcd	0.088 cd	0.056 d	0.132 bc	0.050 d	0.146 b	0.067 d	0.480 a	0.078 d	0.143 b	0.050 d	*	*	*
Kaempferol	0.012 bc	0.002 de	0.012 bcd	0.005 cde	0.014 b	0.003 cde	0.016 b	0.004 cde	0.012 bc	0.001 e	0.031 a	0.002 e		*	*
Myricetin	0.031 abcd	0.027 bcd	0.027 bcd	0.052 a	0.029 bcd	0.017 cd	0.036 abc	0.014 d	0.045 ab	0.011 d	0.020 cd	0.015 d			*
Quercetin	0.025 ab	0.005 d	0.018 c	0.005 d	0.017 c	0.005 d	0.019 bc	0.006 d	0.029 a	0.005 d	0.023 abc	0.005 d		*	*
Rutin	0.179 f	2.809 c	0.041 f	5.101 a	0.062 f	1.299 e	0.091 f	3.390 b	0.053 f	2.478 cd	0.027 f	2.187 d		*	*
Total	9.11 b	4.59 ef	5.56 d	8.35 bc	7.85 c	1.92 g	14.17 a	5.43 d	4.78 de	4.24 ef	5.55 d	3.38 f	*	*	*

L—lighting; S—seasons; a.—acid. Individual phenolic compound content is presented as mg g^{-1} in dry plant matter. Means with different letters (based on heatmapvalues with the same letters on a separate line are marked with the same color) are significantly different at the $p < 0.05$ level by Tukey's honestly significant difference test (*).

3. Discussion

According to various literature sources, phenolic compounds show plasticity in response to light quality, quantity, and duration, allowing plants to adapt to their changes and act as sunscreen, antioxidants, or both. Many studies related to the light quality concern UV-A and blue light as having the most effective impact on phenylpropanoid metabolism than the other light spectrum [6,11,14,19,22,24,29,30]. It was determined that such light stimulates the genes expression belonging to the phenylpropanoid pathway, which is involved in the biosynthesis of phenolic acids and flavonoids mediated by cryptochromes [11,22,24,29,31,32]. In the present study, we used supplemental LEDs of 455 nm, and 400 nm wavelengths, which also could be attributed to the UV-A light spectrum [20,24]. Our data showed that monochromatic supplemental 400 nm LEDs light positively affected the accumulation of total phenolic compounds in lettuce during autumn cultivation but not spring. However, its effect on individual phenolic compounds differed. The significantly higher content of chlorogenic acid and rutin, which were the main part of total phenolic compounds, were determined when such light was applied at nighttime.

Meanwhile, 400 nm LEDs light application at daytime positively affected chichoric acid, epicatechin, and quercetin. O-coumaric acid content increased in both cases. Other authors also reported that a shorter blue or UV-A wavelength enhanced the accumulation of the phenolic compounds in various plants. For example, Lee and coauthors [20] stated that short-term 385 nm UV-A exposure resulted in a significant increase of total phenols, caffeic acid, and kaempferol, but not ferulic acid. Treatments with specific white LEDs light contained a shorter blue wavelength enhanced the accumulation of the individual compounds in butterhead and romaine lettuce cultivars compared to longer ones [30] as well as total phenolics and flavonoids in pak choi [31]. Taulavuori and coauthors [32] showed that violet (420 nm) containing blue (440 nm) light was slightly more effective in the stimulation of flavonoid synthesis in arugula than only blue (450 nm) light. It is known that some flavonoids mostly absorb at 400–430 nm wavelengths light, so it was presumed that shorter blue wavelengths with higher energy could efficiently promote phenolic acid and flavonoid accumulation [30,33].

Meanwhile, supplemental 455 nm LED light applied at nighttime significantly decreased and used at daytime only slightly increased total phenolic compounds content compared to HPS lighting. Such light effect on individual phenolic compounds depended on application time and season. For example, supplemental 455 nm LEDs light at daytime enhanced caffeic acid accumulation in lettuce during autumn cultivation and applied at nighttime during spring cultivation. Meanwhile, chlorogenic acid content increased during spring cultivation when 455 nm LEDs light was used in the daytime and positively affected rosmarinic acid content in all cases. Other authors noticed the positive impact of longer blue wavelengths (450–470 nm) on the accumulation of phenolic compounds in leafy vegetables such as different lettuces varieties, pak choi, Chinese kale, basil, etc. depending on exposure duration till harvest, photoperiod during the daytime, intensity [6,11,14,19,33,34]. Various authors showed that the transcriptional levels of flavonoid biosynthetic genes are strongly affected by the time duration, amount, and blue or UV-A wavelength of light [24,31,33]. According to our results, it can be assumed that 400 nm light stimulates the genes expression belonging to the phenylpropanoid pathway more in comparison to 455 nm. On other hand, our results would suggest that the photoperiod of 455 nm LEDs exposure could be longer and/or higher light intensity for stimulation such genes expression and enhancing phenolic compounds accumulation in lettuce, but further studies are needed.

It is known that green light, the same as blue and UV, is also absorbed by the cryptochromes, although the specific photoreceptor of such light remains to be identified in higher plants. The green light can be more efficiently absorbed by the outer leaves of the canopy and stimulate photosynthesis at lower leaf levels [6,11,24]. However, green LEDs are not extensively applied in commercial plant cultivation due to the inefficiency in converting electricity into photons [8,29,35]. Therefore, there is comparatively little literature on its impacts on plants quality, including phenolic compounds. Few accessible reports showed that green light used as monochromatic or as part of a broader light combination frequently had no effect or reduced accumulation of phenolic compounds and can reverse the positive impact of monochromatic blue light [6,11,14,24,36–39]. The green light had positive results in only a few cases. For example, it enhanced total phenolic and total flavonoids production of *Prunella vulgaris* callus cultures [40]. The significant increase of total phenols in green baby leaf lettuce cultivated in January was found under supplemental 535 nm LEDs and HPS lighting [14]. The present study showed no/or negative effect of green light on total phenolic compounds content in lettuce depending on cultivation season. However, there is a lack of literature data about its impacts on individual phenolic compounds. Our data revealed the positive effect of supplemental green light applied in the daytime on rosmarinic acid during both cultivation seasons. It also enhanced the accumulation of quercetin during spring and p-coumaric acid during autumn cultivation. Meanwhile, the green light at nighttime showed more negative effects than daytime application, except caffeic acid, which significantly increased during autumn application.

However, this study revealed the different effects of green light applied with blue in equal proportions (455 + 530 nm). Such combined lighting used in the daytime during autumn cultivation enhanced the accumulation of total phenolic compounds like chicoric, chlorogenic, and rosmarinic acids, which were mainly in total content. Meanwhile, application at nighttime positively affected total phenolic compounds and the above-mentioned phenolic acids during both cultivation seasons and apigenin during spring and rutin during autumn cultivation. It could show that supplemental blue-green light in a darker period of the day or season was more efficient on phenolic compounds accumulation. Meanwhile, such lighting revealed a more evident positive effect than only blue supplemental LEDs light. According to literature, shorter green light wavelengths less than 530 nm are perceived as part of the cryptochrome and phototropin blue light response. However, longer green light wavelengths about 570 nm are sufficient to antagonise blue light cryptochrome activation [41–43]. In our experiments, we used shorter 530 nm green light, which applied with blue light could stimulate the genes expression belonging to the phenylpropanoid pathway similar to UV-A and blue light and enhanced biosynthesis of phenolic acids and flavonoids in lettuce. We did not find other data concerning the effect of blue/green light on polyphenolic content and composition in leafy vegetables. However, Zheng and coauthors [44] reported that, when dichromatic blue-green light in equal proportion was applied for 4 h in the nighttime, the expression level of several key structural genes among the flavonoid biosynthesis pathway in tea plants decreased compared to the application of monochromatic blue or green light. Therefore, the accumulation of anthocyanin and catechins in such plants under blue-green light was lower than under blue light alone [44].

On the other hand, it was reported that plants interpret decreased blue: green ratios as a shade response, which could act as abiotic stress [42]. It is known that phenolic compounds are produced in plants to overcome potential stressful conditions [45,46]. Therefore, supplemental blue-green light to HPS in which spectrum dominated red-orange-yellow, could change this ratio into a decrease and may act as eustress leading to an increase in phenolic compounds content in lettuce. But further research is required to confirm this assumption.

The present study showed that supplemental 660 nm LEDs light decreased or had no significant effect on total phenolic compounds content compared to HPS lighting. However, some phenolic acid such as caffeic, o-coumaric, and flavonoids epicatechin and rutin significantly increased when the supplemental red light was applied in the daytime during autumn cultivation. According to most studies, red light does not promote phenylpropanoid stimulation of polyphenol biosynthesis, but some cases showed the positive effect of such compounds [2,6,16,23,24]. However, red light's impact on the phenolic accumulation in plants depended on the species and variety, leaf age, duration of exposure, etc. For example, sole red light caused an increase of phenolic compounds in the young red and green Perilla leaves, but not in the mature leaves [47]. The 3-day pre-harvest red (638 nm, 300 μmol m^{-2} s^{-1}) light exposure has a more evident effect on the level of total phenolics in baby leaf lettuce grown in greenhouse conditions in winter [16] and mustard [48]. The higher percentage of red light in red-blue lighting resulted in higher gallic acid and quercetin content in the green basil cultivar, but not in red [49]. The supplemental red light was most effective in enhancing the accumulation of chlorogenic, caffeic, and chicoric acids, rutin, kaempferol, and luteolin in red leaf lettuce [50]. In all of these studies reviewed, red light exposure was longer and more intense during the day. Therefore, it can be assumed that the photoperiod of 660 nm LEDs exposure could be longer and/or higher light intensity for more evidence effect on the increase of polyphenol content.

Generally, in our studies, it was observed that although the variation of the total content of phenolic compounds depending on the lighting and growing season was not large, the content of individual phenolics differed several times. Other authors also determined such trend in lettuce plants [50,51]. It was suggested that it is probably related with enhanced activity of the phenylpropanoid pathway, resulting in an increase of intermediate and final products of this branched metabolic pathway. For example, chlorogenic acid characterizes a typical compound synthesized within the phenylpropanoid pathway and relates to p-coumaric acid as an intermediate product [51].

Few studies have described the seasonal effect on phenolic compounds and their individual composition in leafy vegetables, suggesting that variation in phenolic compounds is more dependent on growing conditions and cultivar [14,16,27,52–54]. For example, it was observed that differences among five lettuce cultivars appear to have a more significant impact on phenolic compounds than environmental variation during the growing season [52]. The best light combination for increasing phenolic compounds content in Lamb's lettuce grown in the greenhouse was 70R/30B in autumn and 50R/50B in winter cultivation [27]. Samuoliene and coauthors [14,16] showed that total phenols content in various baby leaf lettuces mainly increased in the darker months—November and January—when the supplemental blue, green or red light was applied together with HPS lighting. Marin and coauthors [53] observed that the increase in temperature and radiation from February to May promoted the increase in the content of phenolic acids and flavonoids and showed the seasonal variations of individual phenolic compounds. Meanwhile, Lee and coauthors [54] reported that analysed phenolic acids in red Chinese cabbages increased in autumn and flavonols in spring cultivation. The present study showed a significant seasonality effect, but it depends on supplemental LEDs light application time. The application of LEDs during the daytime resulted in higher total phenolic compounds content during autumn cultivation. However, individual phenolic compounds such as gallic, protocatechuic, p-coumaric, rosmarinic acids, myricetin, apigenin, and kaempferol content were higher during spring cultivation. Meanwhile, supplemental LEDs application in the nighttime increased analysed phenolic compounds content during spring cultivation, except rutin. Generally, our results confirm other authors observations that increased radiation in spring increasing phenolics content maybe due to higher stimulation the genes expression of phenylpropanoid pathway [33,53,54]. That suggests different lighting strategies for increasing phenolic compounds content in lettuces during different growing seasons, but further studies are need for better understanding of the regulation phenolics compounds synthesis during different growing seasons

4. Materials and Methods

4.1. Growth Conditions

Pot experiments were performed at the greenhouse of the Institute of Horticulture, Lithuanian Research Centre of Agricultural and Forestry (lat. 55° N), during the autumn and spring periods. Seeds of lettuce (*Lactuca sativa* L. cv. Little Gem; CN Seeds, UK) were sown in cell trays (70 mL cell volume, three seeds per tray, 54 trays in one plastic vessel) containing peat substrate (Terraerden, Rucava, Latvia) with NPK (100–160; 110–180; 120–200 mg L^{-1}) and microelements Mn, Cu, Mo, B, Zn and Fe (pH H_2O 5.5–6.5; electrical conductivity (EC) ms cm^{-1} < 1.10). Seedlings with one true leaf (BBCH 11–12) were transplanted into pots 500 mL, filled with the same peat substrate. Plants were watered when needed, maintaining a similar substrate humidity. From the 10th day after transplanting (BBCH 14–15), plants were watered with 50 mL of nutrient solution for plant (10 mL NPK 3-1-3 (Terra grow, Plagron, Netherlands) in 5 L water) two times a week. The day/night temperatures of 23 $\pm$ 2/16 $\pm$ 2 °C, and 16-h photoperiod and relative air humidity of 70 $\pm$ 10% were maintained. Plants were grown under daylight with supplementary lighting provided by standard high-pressure sodium lamps (HPS) (SON-T Agro, 400 W, Philips, Eindhoven, The Netherlands) 16-h photoperiod. The generated photosynthetic photon flux density (PPFD) of HPS lamps at plant level was about 90 $\pm$ 10 µmol m^{-2} s^{-1}. During the experiments, the weekly-average solar radiation inside the greenhouse ranged from 20 to 80 µmol m^{-2} s^{-1} in November–December and from 150 to 230 µmol m^{-2} s^{-1} in March–April. At thirty days after germination, lighting experiments began (see Section 4.2). At the end of the experiments, plants were harvested just above the substrate level. Samples of randomly selected twelve fully developed lettuce plants per treatment were used for phytochemical analysis.

4.2. Lighting Treatments

At the lettuce pre-harvest stage of 5 days, the HPS lamps were supplemented by LEDs lamps (4h photoperiod) (Figures 1 and 2). Different LEDs lamps (Vegetal Grow Development, France) contained diodes with the peak wavelength of UV-A (400 nm), blue (455 nm), green (530 nm), blue + green (455 + 530 nm) and red (660 nm). Two lighting experiments (EXP1 and EXP2) with replication were carried out during the autumn and spring periods. EXP1 – HPS + LEDs lamps lighting was from 06 till 10 AM where HPS lamps generated 90 $\pm$ 10 µmol m^{-2} s^{-1} and LEDs generated 50 µmol m^{-2} s^{-1} (Figure 1). Later, only HPS lighting was from 10 AM till 10 PM. EXP2—HPS lighting was from 06 AM till 10 PM, and LEDs lamps lighting was from 10 PM till 02 AM (Figure 2). A photometer RF-100 with head G.PAR-100 was used to measure PPFD (Sonopan, Bialystok, Poland).

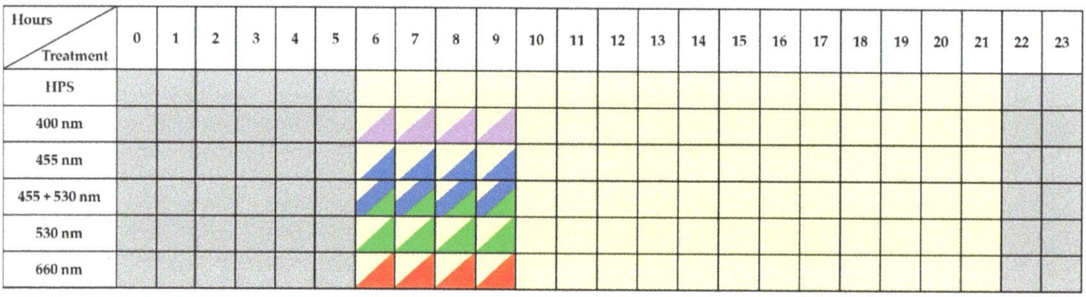

Figure 1. The lighting scheme of 1st experiment (EXP1).

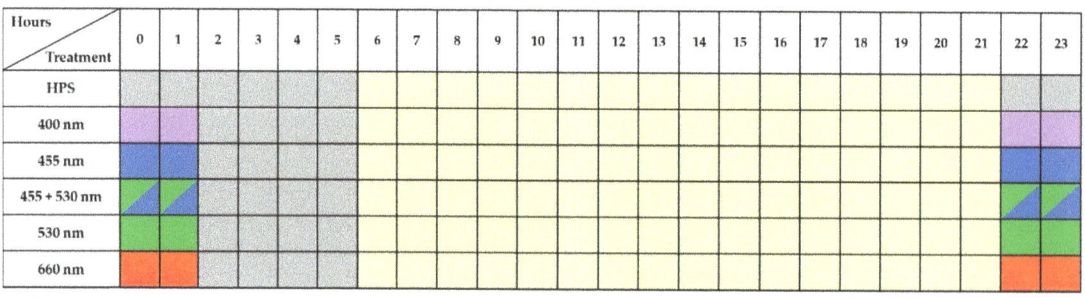

Figure 2. The lighting scheme of the 2nd experiment (EXP2).

4.3. Determination of Individual Phenolic Compounds

The fresh plant material was immediately frozen in liquid N_2 and lyophilised for in-dividual phenolic compound analysis. It was calculated that the average content of dry matter in lettuce was about 4.5% during both growing seasons. Individual phenolic compounds were analysed by high-performance liquid chromatography (HPLC) on a NUCLEODUR Sphinx RP column (5 μm particle size, 150 × 4.6 mm) (Macherey-Nagel GmbH & Co KG, Düren Germany). For phenolic compounds extraction, 100 mg of lyophilised plant material was grounded with 80% ice-cold methanol (Sigma-Aldrich, St. Louis, MO, USA) and transferred to a 15 mL polypropylene conical centrifuge tube (Labbox Labware S.L., Barcelona, Spain). The extract was incubated at 4 °C for 24 h. Samples were centrifuged (Hermle Z 300 K, Hermle Labortechnik, Wehingen, Germany) at a relative centrifugal force of 4000 rpm min^{-1} for 10 min at room temperature. The supernatant was filtered through a 70 mm qualitative filter paper (Frisenette ApS, Knebel, Denmark). Before the HPLC analyses, samples were filtrated through a 13 mm and 0.22 μm nylon syringe filter (BGB Analytik AG, Böckten, Switzerland). The HPLC 10A system (Shimadzu, Kyoto, Japan) equipped with a diode array (SPD-M 10A VP) detector was used for analysis. Peaks were detected at 280 nm. The mobile phase consisted of A (100% acetonitrile, (Sigma-Aldrich, St. Louis, MO, USA) and B (1% acetic acid, Supelco, Bellefonte, PA, USA). Binary gradient: 0 min; 95% B, 25 min; 70% B, 25–30 min; 5% B, 30–35; 5% B; 35–37 min; 95% B, and 37–40 min; 95% B, flow rate 1 mL min^{-1}. The results are expressed as an average of analytical measurements of three biological samples from homogenized plant material in mg g^{-1} in the dry mass of plants. The contents of rutin (rutin trihydrate, Supelco), myricetin, chicoric acid, ferulic acid (*trans*-Ferulic acid), rosmarinic acid, quercetin and protocatechuic acid (all purchased from Merck KGaA, Darmstadt, Germany), caffeic acid, p-coumaric acid (*trans*-p-Coumaric acid), o-coumaric acid (*trans*-2-Hydroxycinnamic acid), m-coumaric acid (*trans*-3-Hydroxycinnamic acid), epicatechin, chlorogenic acid, kaempferol, gallic acid (gallic acid monohydrate) (all purchased from Sigma-Aldric), and apigenin (LGC Standards Ltd., LGC, Teddington, UK) are expressed as mg g^{-1} in the dry matter of plants.

4.4. Statistical Analysis

Statistical analysis was performed using Microsoft Excel 2016 and Addinsoft XLSTAT 2022 XLSTAT statistical and data analysis (Long Island, NY, USA). Two-way analysis of variance (ANOVA) followed by Tukey's honestly significant difference test ($p < 0.05$) for multiple comparisons was used to evaluate differences between means ($n = 3$) of measurements.

5. Conclusions

This study demonstrated that even short-term supplemental LEDs preharvest lighting affected the accumulation of total and individual phenolic compounds in lettuces. However, our results suggest different LEDs application strategies for increasing their content during different growing seasons and times of the day. The most pronounced positive effect on total phenolic compounds revealed supplemental 400 and 455 + 530 nm LEDs lighting, except its application during the daytime during spring cultivation, when all supplemental LEDs light had no impact on such compound. Supplemental 400 nm LEDS applied in the daytime increased primary phenolic compounds such as chlorogenic acid during spring and chicoric acid during autumn cultivation. Meanwhile, 400 nm LEDs used in nighttime enhanced chlorogenic acid accumulation and rutin during autumn but were not effective during spring cultivation. Chicoric and chlorogenic acid significantly increased under supplemental 455 + 530 nm LEDs applied at daytime in autumn and used at nighttime—in spring. Supplemental LEDs application in the nighttime resulted in higher analysed phenolic compounds content during spring cultivation, except rutin. When applied in the daytime, higher total phenolic compounds, chicoric and chlorogenic acid content were determined during autumn cultivation. A review of our and the literature data suggests that further research is required to clarify the impact of more prolonged and intense supplemental LEDs to HPS lighting exposure on various phenolic compounds accumulation and biosynthetic pathways in lettuce.

Supplementary Materials: The following are available online at https://www.mdpi.com/article/10.3390/plants11091123/s1. Table S1. Effect of short-term daytime supplemental LEDs to HPS lighting on phenolic compounds content in lettuce cultivated in a greenhouse. Table S2. Effect of different seasons on phenolic compounds content in lettuce cultivated in a greenhouse under short-term daytime supplemental LEDs to HPS. Table S3. Effect of short-term nighttime supplemental LEDs to HPS lighting on phenolic compounds content in lettuce cultivated in a greenhouse. Table S4. Effect of different seasons on phenolic compounds content in lettuce cultivated in a greenhouse under short-term nighttime supplemental LEDs to HPS.

Author Contributions: Conceptualisation, A.B., V.V.-K., N.R., A.V. (Akvilė Viršilė), A.V. (Alma Valiuškaitė) and G.S.; methodology, A.B., V.V.-K., J.M., N.R., A.V. (Akvilė Viršilė), A.V. (Alma Valiuškaitė) and G.S.; software, V.V.-K.; validation, A.B., V.V.-K., N.R. and G.S.; performing the experiment, collecting the samples for analysis, R.S., J.M., K.L., L.D., S.C. and A.K.; formal analysis, V.V.-K.; investigation, A.B., V.V.-K., N.R., A.V. (Akvilė Viršilė), A.V. (Alma Valiuškaitė) and G.S.; writing-original draft preparation, A.B. and V.V.-K.; writing-review and editing, N.R., A.V. (Akvilė Viršilė), A.V. (Alma Valiuškaitė), J.M. and G.S.; visualisation, A.B. and V.V.-K. All authors have read and agreed to the published version of the manuscript.

Funding: This project has received funding from European Regional Development Fund (project No 2.2-LMT-K-718-03-0035) under grant agreement with the Research Council of Lithuania (LMTLT).

Institutional Review Board Statement: Not applicable.

Informed Consent Statement: Not applicable.

Data Availability Statement: Not applicable.

Conflicts of Interest: The authors declare no conflict of interest.

References

1. Rouphael, Y.; Kyriacou, M.C.; Petropoulos, S.A.; De Pascale, S.; Colla, G. Improving Vegetable Quality in Controlled Environments. *Sci. Hortic.* **2018**, *234*, 275–289. [CrossRef]
2. Thoma, F.; Somborn-Schulz, A.; Schlehuber, D.; Keuter, V.; Deerberg, G. Effects of Light on Secondary Metabolites in Selected Leafy Greens: A Review. *Front. Plant Sci.* **2020**, *11*, 497. [CrossRef] [PubMed]
3. Appolloni, E.; Pennisi, G.; Zauli, I.; Carotti, L.; Paucek, I.; Quaini, S.; Orsini, F.; Gianquinto, G. Beyond Vegetables: Effects of Indoor LED Light on Specialized Metabolite Biosynthesis in Medicinal and Aromatic Plants, Edible Flowers, and Microgreens. *J. Sci. Food Agric.* **2022**, *102*, 472–487. [CrossRef] [PubMed]
4. Cory, H.; Passarelli, S.; Szeto, J.; Tamez, M.; Mattei, J. The Role of Polyphenols in Human Health and Food Systems: A Mini-Review. *Front. Nutr.* **2018**, *5*, 87. [CrossRef]
5. Di Lorenzo, C.; Colombo, F.; Biella, S.; Stockley, C.; Restani, P. Polyphenols and Human Health: The Role of Bioavailability. *Nutrients* **2021**, *13*, 273. [CrossRef]
6. Wong, C.E.; Teo, Z.W.N.; Shen, L.; Yu, H. Seeing the Lights for Leafy Greens in Indoor Vertical Farming. *Trends Food Sci. Technol.* **2020**, *106*, 48–63. [CrossRef]
7. Viršilė, A.; Brazaitytė, A.; Jankauskienė, J.; Miliauskienė, J.; Vaštakaitė, V.; Odminytė, I.; Novičkovas, A.; Samuolienė, G. Pre-Harvest LED Lighting Strategies for Reduced Nitrate Contents in Leafy Vegetables. *Zemdirb.-Agric.* **2018**, *105*, 249–256. [CrossRef]
8. Mitchell, C.A.; Dzakovich, M.P.; Gomez, C.; Lopez, R.; Burr, J.F.; Hernández, R.; Kubota, C.; Currey, C.J.; Meng, Q.; Runkle, E.S.; et al. Light-Emitting Diodes in Horticulture. In *Horticultural Reviews: Volume 43*; Janick, J., Ed.; John Wiley & Sons, Inc.: Hoboken, NJ, USA, 2015; pp. 1–88, ISBN 9781119107781.
9. Ouzounis, T.; Rosenqvist, E.; Ottosen, C.-O. Spectral Effects of Artificial Light on Plant Physiology and Secondary Metabolism: A Review. *HortScience* **2015**, *50*, 1128–1135. [CrossRef]
10. Bantis, F.; Smirnakou, S.; Ouzounis, T.; Koukounaras, A.; Ntagkas, N.; Radoglou, K. Current Status and Recent Achievements in the Field of Horticulture with the Use of Light-Emitting Diodes (LEDs). *Sci. Hortic.* **2018**, *235*, 437–451. [CrossRef]
11. Zhang, X.; Bian, Z.; Yuan, X.; Chen, X.; Lu, C. A Review on the Effects of Light-Emitting Diode (LED) Light on the Nutrients of Sprouts and Microgreens. *Trends Food Sci. Technol.* **2020**, *99*, 203–216. [CrossRef]
12. Azad, M.O.K.; Kjaer, K.H.; Adnan, M.; Naznin, M.T.; Lim, J.D.; Sung, I.J.; Park, C.H.; Lim, Y.S. The Evaluation of Growth Performance, Photosynthetic Capacity, and Primary and Secondary Metabolite Content of Leaf Lettuce Grown under Limited Irradiation of Blue and Red LED Light in an Urban Plant Factory. *Agriculture* **2020**, *10*, 28. [CrossRef]

13. Rahman, M.M.; Field, D.L.; Ahmed, S.M.; Hasan, M.T.; Basher, M.K.; Alameh, K. LED Illumination for High-Quality High-Yield Crop Growth in Protected Cropping Environments. *Plants* **2021**, *10*, 2470. [CrossRef] [PubMed]
14. Samuolienė, G.; Sirtautas, R.; Brazaitytė, A.; Duchovskis, P. LED Lighting and Seasonality Effects Antioxidant Properties of Baby Leaf Lettuce. *Food Chem.* **2012**, *134*, 1494–1499. [CrossRef] [PubMed]
15. Samuolienė, G.; Urbonavičiūtė, A.; Duchovskis, P.; Bliznikas, Z.; Vitta, P.; Žukauskas, A. Decrease in Nitrate Concentration in Leafy Vegetables Under a Solid-State Illuminator. *HortScience* **2009**, *44*, 1857–1860. [CrossRef]
16. Samuolienė, G.; Sirtautas, R.; Brazaitytė, A.; Viršilė, A.; Duchovskis, P. Supplementary Red-LED Lighting and the Changes in Phytochemical Content of Two Baby Leaf Lettuce Varieties During Three Seasons. *J. Food. Agric. Environ.* **2012**, *10*, 701–706. [CrossRef]
17. Zhou, W.L.; Liu, W.K.; Yang, Q.C. Quality Changes in Hydroponic Lettuce Grown under Pre-Harvest Short-Duration Continuous Light of Different Intensities. *J. Hortic. Sci. Biotechnol.* **2012**, *87*, 429–434. [CrossRef]
18. Wanlai, Z.; Wenke, L.; Qichang, Y. Reducing Nitrate Content in Lettuce by Pre-Harvest Continuous Light Delivered by Red and Blue Light-Emitting Diodes. *J. Plant Nutr.* **2013**, *36*, 481–490. [CrossRef]
19. Ouzounis, T.; Parjikolaei, R.B.; Fretté, X.; Rosenqvist, E.; Ottosen, C.-O. Predawn and High Intensity Application of Supplemental Blue Light Decreases the Quantum Yield of PSII and Enhances the Amount of Phenolic Acids, Flavonoids, and Pigments in Lactuca Sativa. *Front. Plant Sci.* **2015**, *6*, 19. [CrossRef]
20. Lee, J.-H.; Oh, M.-M.; Son, K.-H. Short-Term Ultraviolet (UV)-A Light-Emitting Diode (LED) Radiation Improves Biomass and Bioactive Compounds of Kale. *Front. Plant Sci.* **2019**, *10*, 1042. [CrossRef]
21. Kopsell, D.A.; Sams, C.E. Increases in Shoot Tissue Pigments, Glucosinolates, and Mineral Elements in Sprouting Broccoli after Exposure to Short-Duration Blue Light from Light Emitting Diodes. *J. Am. Soc. Hortic. Sci.* **2013**, *138*, 31–37. [CrossRef]
22. Jiang, H.; Li, X.; Tian, J.; Liu, H. Pre-Harvest Supplemental Blue Light Enhanced Antioxidant Activity of Flower Stalk in Chinese Kale during Storage. *Plants* **2021**, *10*, 1177. [CrossRef] [PubMed]
23. Brazaitytė, A.; Sakalauskienė, S.; Viršilė, A.; Jankauskienė, J.; Samuolienė, G.; Sirtautas, R.; Vaštakaitė, V.; Miliauskienė, J.; Duchovskis, P.; Novičkovas, A.; et al. The Effect of Short-Term Red Lighting on *Brassicaceae* Microgreens Grown Indoors. *Acta Hortic.* **2016**, *1123*, 177–184. [CrossRef]
24. Landi, M.; Zivcak, M.; Sytar, O.; Brestic, M.; Allakhverdiev, S.I. Plasticity of Photosynthetic Processes and the Accumulation of Secondary Metabolites in Plants in Response to Monochromatic Light Environments: A Review. *Biochim. Biophys. Acta Bioenerg.* **2020**, *1861*, 148131. [CrossRef] [PubMed]
25. Soto, V.C.; González, R.E.; Galmarini, C.R. Bioactive Compounds in Vegetables, Is There Consistency in the Published Information? A Systematic Review. *J. Hortic. Sci. Biotechnol.* **2021**, *96*, 570–587. [CrossRef]
26. Wojciechowska, R.; Kołton, A.; Długosz-Grochowska, O.; Żupnik, M.; Grzesiak, W. The Effect of LED Lighting on Photosynthetic Parameters and Weight of Lamb's Lettuce (*Valerianella Locusta*). *Folia Hortic.* **2013**, *25*, 41–47. [CrossRef]
27. Długosz-Grochowska, O.; Kołton, A.; Wojciechowska, R. Modifying Folate and Polyphenol Concentrations in Lamb's Lettuce by the Use of LED Supplemental Lighting during Cultivation in Greenhouses. *J. Funct. Foods* **2016**, *26*, 228–237. [CrossRef]
28. Kim, M.J.; Moon, Y.; Tou, J.C.; Mou, B.; Waterland, N.L. Nutritional Value, Bioactive Compounds and Health Benefits of Lettuce (*Lactuca sativa* L.). *J. Food Compos. Anal.* **2016**, *49*, 19–34. [CrossRef]
29. Thwe, A.A.; Kim, Y.B.; Li, X.; Seo, J.M.; Kim, S.-J.; Suzuki, T.; Chung, S.-O.; Park, S.U. Effects of Light-Emitting Diodes on Expression of Phenylpropanoid Biosynthetic Genes and Accumulation of Phenylpropanoids in Fagopyrum Tataricum Sprouts. *J. Agric. Food Chem.* **2014**, *62*, 4839–4845. [CrossRef]
30. Nguyen, T.K.L.; Cho, K.M.; Lee, H.Y.; Cho, D.Y.; Lee, G.O.; Jang, S.N.; Lee, Y.; Kim, D.; Son, K.-H. Effects of White LED Lighting with Specific Shorter Blue and/or Green Wavelength on the Growth and Quality of Two Lettuce Cultivars in a Vertical Farming System. *Agronomy* **2021**, *11*, 2111. [CrossRef]
31. Mao, P.; Duan, F.; Zheng, Y.; Yang, Q. Blue and UV-A Light Wavelengths Positively Affected Accumulation Profiles of Healthy Compounds in Pak-choi. *J. Sci. Food Agric.* **2021**, *101*, 1676–1684. [CrossRef]
32. Taulavuori, K.; Pyysalo, A.; Taulavuori, E.; Julkunen-Tiitto, R. Responses of Phenolic Acid and Flavonoid Synthesis to Blue and Blue-Violet Light Depends on Plant Species. *Environ. Exp. Bot.* **2018**, *150*, 183–187. [CrossRef]
33. Gitelson, A.; Chivkunova, O.; Zhigalova, T.; Solovchenko, A. In Situ Optical Properties of Foliar Flavonoids: Implication for Non-Destructive Estimation of Flavonoid Content. *J. Plant Physiol.* **2017**, *218*, 258–264. [CrossRef] [PubMed]
34. Zheng, Y.; Zhang, Y.; Liu, H.; Li, Y.; Liu, Y.; Hao, Y.; Lei, B. Supplemental Blue Light Increases Growth and Quality of Greenhouse Pak Choi Depending on Cultivar and Supplemental Light Intensity. *J. Integr. Agric.* **2018**, *17*, 2245–2256. [CrossRef]
35. Kelly, N.; Vaštakaitė-Kairienė, V.; Runkle, E.S. Indoor lighting effects on plant nutritional compounds. In *Plant Factory Basics, Applications and Advances*; Kozai, T., Niu, G., Masabni, J., Eds.; Accademic Press: Cambridge, MA, USA, 2022; pp. 329–349.
36. Vaštakaitė, V.; Viršilė, A.; Brazaitytė, A.; Samuolienė, G.; Jankauskienė, J.; Novičkovas, A.; Duchovskis, P. Pulsed Light-Emitting Diodes for a Higher Phytochemical Level in Microgreens. *J. Agric. Food Chem.* **2017**, *65*, 6529–6534. [CrossRef] [PubMed]
37. Dou, H.; Niu, G.; Gu, M. Photosynthesis, Morphology, Yield, and Phytochemical Accumulation in Basil Plants Influenced by Substituting Green Light for Partial Red and/or Blue Light. *Horts* **2019**, *54*, 1769–1776. [CrossRef]
38. Son, K.-H.; Oh, M.-M. Growth, Photosynthetic and Antioxidant Parameters of Two Lettuce Cultivars as Affected by Red, Green, and Blue Light-Emitting Diodes. *Hortic. Environ. Biotechnol.* **2015**, *56*, 639–653. [CrossRef]

39. Zhang, X.; Bian, Z.; Li, S.; Chen, X.; Lu, C. Comparative Analysis of Phenolic Compound Profiles, Antioxidant Capacities, and Expressions of Phenolic Biosynthesis-Related Genes in Soybean Microgreens Grown under Different Light Spectra. *J. Agric. Food Chem.* **2019**, *67*, 13577–13588. [CrossRef]
40. Fazal, H.; Abbasi, B.H.; Ahmad, N.; Ali, S.S.; Akbar, F.; Kanwal, F. Correlation of Different Spectral Lights with Biomass Accumulation and Production of Antioxidant Secondary Metabolites in Callus Cultures of Medicinally Important Prunella *Vulgaris* L. *J. Photochem. Photobiol. B Biol.* **2016**, *159*, 1–7. [CrossRef]
41. Sellaro, R.; Crepy, M.; Trupkin, S.A.; Karayekov, E.; Buchovsky, A.S.; Rossi, C.; Casal, J.J. Cryptochrome as a Sensor of the Blue/Green Ratio of Natural Radiation in Arabidopsis. *Plant Physiol.* **2010**, *154*, 401–409. [CrossRef]
42. Battle, M.W.; Jones, M.A. Cryptochromes Integrate Green Light Signals into the Circadian System. *Plant Cell Environ.* **2020**, *43*, 16–27. [CrossRef]
43. Smith, H.L.; McAusland, L.; Murchie, E.H. Don't Ignore the Green Light: Exploring Diverse Roles in Plant Processes. *J. Exp. Bot.* **2017**, *68*, 2099–2110. [CrossRef] [PubMed]
44. Zheng, C.; Ma, J.-Q.; Ma, C.-L.; Shen, S.-Y.; Liu, Y.-F.; Chen, L. Regulation of Growth and Flavonoid Formation of Tea Plants (*Camellia sinensis*) by Blue and Green Light. *J. Agric. Food Chem.* **2019**, *67*, 2408–2419. [CrossRef] [PubMed]
45. Sharma, A.; Shahzad, B.; Rehman, A.; Bhardwaj, R.; Landi, M.; Zheng, B. Response of Phenylpropanoid Pathway and the Role of Polyphenols in Plants under Abiotic Stress. *Molecules* **2019**, *24*, 2452. [CrossRef] [PubMed]
46. Šamec, D.; Karalija, E.; Šola, I.; Vujčić Bok, V.; Salopek-Sondi, B. The Role of Polyphenols in Abiotic Stress Response: The Influence of Molecular Structure. *Plants* **2021**, *10*, 118. [CrossRef] [PubMed]
47. Viršilė, A.; Brazaitytė, A.; Sirtautas, R.; Duchovskis, P. Light Spectral Effects on Phenolic Compounds in *Perilla frutescens* Leaves as Related to the Leaf Age, Color and Duration of Exposure. *Acta Hortic.* **2017**, *1170*, 981–988. [CrossRef]
48. Bliznikas, Z.; Zukauskas, A.; Samuolienė, G.; Viršilė, A.; Brazaitytė, A.; Jankauskienė, J.; Duchovskis, P.; Novičkovas, A. Effect of Supplementary Pre-harvest LED Lighting on the Antioxidant and Nutritional Properties of Green Vegetables. *Acta Hortic.* **2012**, *939*, 85–91. [CrossRef]
49. Lobiuc, A.; Vasilache, V.; Oroian, M.; Stoleru, T.; Burducea, M.; Pintilie, O.; Zamfirache, M.-M. Blue and Red LED Illumination Improves Growth and Bioactive Compounds Contents in Acyanic and Cyanic Ocimum *Basilicum* L. Microgreens. *Molecules* **2017**, *22*, 2111. [CrossRef]
50. Lee, M.; Xu, J.; Wang, W.; Rajashekar, C.B. The Effect of Supplemental Blue, Red and Far-Red Light on the Growth and the Nutritional Quality of Red and Green Leaf Lettuce. *Am. J. Plant Sci.* **2019**, *10*, 2219–2235. [CrossRef]
51. Sytar, O.; Zivcak, M.; Bruckova, K.; Brestic, M.; Hemmerich, I.; Rauh, C.; Simko, I. Shift in Accumulation of Flavonoids and Phenolic Acids in Lettuce Attributable to Changes in Ultraviolet Radiation and Temperature. *Sci. Hortic.* **2018**, *239*, 193–204. [CrossRef]
52. Bunning, M.L.; Kendall, P.A.; Stone, M.B.; Stonaker, F.H.; Stushnoff, C. Effects of Seasonal Variation on Sensory Properties and Total Phenolic Content of 5 Lettuce Cultivars. *J. Food Sci.* **2010**, *75*, S156–S161. [CrossRef]
53. Marin, A.; Ferreres, F.; Barberá, G.G.; Gil, M.I. Weather Variability Influences Color and Phenolic Content of Pigmented Baby Leaf Lettuces throughout the Season. *J. Agric. Food Chem.* **2015**, *63*, 1673–1681. [CrossRef] [PubMed]
54. Lee, H.; Oh, I.-N.; Kim, J.; Jung, D.; Cuong, N.P.; Kim, Y.; Lee, J.; Kwon, O.; Park, S.U.; Lim, Y.; et al. Phenolic Compound Profiles and Their Seasonal Variations in New Red-Phenotype Head-Forming Chinese Cabbages. *LWT* **2018**, *90*, 433–439. [CrossRef]

plants

MDPI

Article

Light Quality Modulates Photosynthesis and Antioxidant Properties of *B. vulgaris* L. Plants from Seeds Irradiated with High-Energy Heavy Ions: Implications for Cultivation in Space

Ermenegilda Vitale [1,2], Luigi Gennaro Izzo [2], Chiara Amitrano [2], Violeta Velikova [3], Tsonko Tsonev [3], Palma Simoniello [4], Veronica De Micco [2] and Carmen Arena [1,5,*]

1 Department of Biology, University of Naples Federico II, Via Cinthia, 80126 Naples, Italy; ermenegilda.vitale@unina.it
2 Department of Agricultural Sciences, University of Naples Federico II, Via Università 100, 80055 Portici, Italy; luigigennaro.izzo@unina.it (L.G.I.); chiara.amitrano@unina.it (C.A.); demicco@unina.it (V.D.M.)
3 Institute of Plant Physiology and Genetics, Bulgarian Academy of Sciences, Acad. G. Bonchev Str., bl. 21, 1113 Sofia, Bulgaria; violet@bio21.bas.bg (V.V.); ttsonev@bio21.bas.bg (T.T.)
4 Department of Science and Technology, Parthenope University of Naples, Via Acton 38, 80133 Naples, Italy; palma.simoniello@uniparthenope.it
5 BAT Center—Center for Studies on Bioinspired Agro-Environmental Technology, 80055 Portici, Italy
* Correspondence: c.arena@unina.it

Abstract: *Beta vulgaris* L. is a crop selected for cultivation in Space for its nutritional properties. However, exposure to ionizing radiation (IR) can alter plant photosynthetic performance and phytochemical production in the extraterrestrial environment. This study investigated if plant growth under different light quality regimes (FL—white fluorescent; RGB—red–green–blue; RB—red–blue) modifies the photosynthetic behavior and bioactive compound synthesis of plants sprouted by dry seeds irradiated with carbon or titanium high-energy ions. The study evidenced that: (i) the plant response depends on the type of heavyion; (ii) control and C-ion-irradiated plants were similar for photosynthetic pigment content and PSII photochemical efficiency, regardless of the LQ regime; (iii) under FL, net photosynthesis (A_N) and water use efficiency (iWUE) declined in C- and Ti-ion plants compared to control, while the growth of irradiated plants under RGB and RB regimes offset these differences; (iv) the interaction Ti-ion× RB improved iWUE, and stimulated the production of pigments, carbohydrates, and antioxidants. The overall results highlighted that the cultivation of irradiated plants under specific LQ regimes effectively regulates photosynthesis and bioactive compound amounts in leaf edible tissues. In particular, the interaction Ti-ion × RB improved iWUE and increased pigments, carbohydrates, and antioxidant content.

Keywords: antioxidants; *Beta vulgaris* L.; ionizing radiation; light quality; photosynthesis; Space Closed Ecosystem

Citation: Vitale, E.; Izzo, L.G.; Amitrano, C.; Velikova, V.; Tsonev, T.; Simoniello, P.; De Micco, V.; Arena, C. Light Quality Modulates Photosynthesis and Antioxidant Properties of *B. vulgaris* L. Plants from Seeds Irradiated with High-Energy Heavy Ions: Implications for Cultivation in Space. *Plants* **2022**, *11*, 1816. https://doi.org/10.3390/plants11141816

Academic Editors: Hail Rihan and Mick Fuller

Received: 24 June 2022
Accepted: 7 July 2022
Published: 10 July 2022

1. Introduction

The realization of Bioregenerative Life Support Systems (BLSSs) is crucial in considering future long-term human-crewed missions in Space. Transit vehicles, space stations, and platforms on Moon and Mars will include self-sustaining artificial eco-systems based on the balance between heterotrophs (humans and microorganisms) and autotrophs (plants or algae). In particular, higher plants significantly contribute to re-storing the resources in closed environments, regenerating and purifying air through CO_2 absorption and O_2 evolution and transpiration, as well as producing fresh food supplies for the crew [1–4].

Space is a harsh environment compared to Earth. Many factors may constrain the plant's survival in the extraterrestrial environment, including altered gravity, the interaction between microgravity and fluid dynamics, ionizing radiation (IR), and modified pressure and temperature conditions. Among space factors, ionizing radiation represents the main hazard for the survival of life forms, including plants, in exploratory-class missions [5].

Understanding the effects of IR on photosynthetic apparatus and, in general, on plant metabolism is a prerequisite for cultivating plants in Space. IR may affect the photosynthetic process at different levels: from molecular, impacting light-harvesting complexes, reaction centers, and electron transport carriers, to physiological level by affecting primary and secondary photosynthetic metabolism, also through anatomical changes of leaf structure [6–9]. In addition, the radio-induced stress in plants triggers the production of a large variety of antioxidant compounds which are engaged in the detoxification of reactive oxygen species (ROS) and, at the same time, enriches the nutritional properties of plant tissues [4,10–12].

Generally, plant responses to IR depend on several variables, including species, phenological stage at the time of irradiation, dose, and radiation quality [5,6,13]. The space radiation environment consists of a wide variety of ion species with a continuous range of energies. The principal galactic cosmic rays (GCR) include high-energy protons, alpha particles, and heavy ions (HZE—high-energy nuclei component). Therefore, testing plant response to specific ions at proper acute doses is a vital prerequisite to assessing plant radiosensitivity and evaluating the suitability of different crops for cultivation in Space.

Carbon (C) and titanium (Ti) are among the heavy ions considered representative of HZE and are often used to simulate the GRC spectrum in ground-based experiments [14,15]. However, very little is known about titanium conversely to carbon ions. Early studies on animal models evidence that Ti-ions induce oxidative stress and genomic alterations associated with several health risks [16–18]. In plants, Ti-ions have been reported to improve starch mobilization towards actively growing tissues of eye bean seedlings and stimulate the production of antioxidants [19].

Therefore, the defining agricultural practices, as well as micro-environmental parameters, are essential for the selection of suitable crops for space farming. For example, plants have different requirements for light intensity, quality, and duration [3,20]. In particular, the light spectral composition affects not only germination [21], plant architecture, and leaf anatomy [22–24] but also physiological processes [25], such as stomatal opening regulation [26], photosynthesis [27–29], pigment synthesis [30,31], and ultimately biomass production [20,32–34]. Furthermore, specific light quality treatments during growth may also stimulate the resistance to diseases and abiotic stress (high temperature, nutritional deficiency, and heavy metals) [27,35–37], improving the synthesis of antioxidants [38] which, in turn, can enhance the nutritional quality of crops [39].

Based on this evidence, the modulation of the light spectrum is a promising tool for improving plant productivity in space farming [40–43], also representing a means to face the constraints of the space environment.

From this perspective, studying the interaction between space IR and light quality (LQ) is gaining interest in space research. Recent studies on crops evidenced that the interplay LQ/IR may elicit essential plant traits, such as dwarf growth, increased photosynthesis and nutritional value [44,45]. The present study aimed to deepen the knowledge of the interaction between high-LET (Linear Energy Transfer) ionizing radiation and LQ on chard (*B. vulgaris* L. var. cicla) plants focusing specifically on the photosynthetic process to assess the suitability of this species to be cultivated in Bioregenerative Life Support Systems (BLSSs). Chard was chosen for this study because it is considered a functional food [46], and for the high nutritional value of its leaves, rich in healthy secondary metabolites. Furthermore, its compact size and the great amount of edible biomass make it suitable for Space cultivation [3].

The specific aims of this study were to investigate: (i) how the exposure of chard seeds C and Ti heavy ions, representative particles of the galactic cosmic rays, may affect the photosynthetic metabolism of chard; (ii) how plant development under specific LQ regimes may modify the photosynthetic response to C and Ti irradiation; (iii) if and at what level the interaction between IR and specific LQ treatments may promote the production of functional metabolites which are beneficial as a supplement for the astronauts' diet.

2. Results

2.1. Germination and Plant Biomass

Irradiation with C- and Ti-ions caused a significant increase in GP (100%) under FL light compared to not-irradiated control. RB light also determined a significant increase in GP after irradiation with C- (100%) and Ti-ions (75%). The same tendency to increase GP after irradiation was found under RGB light, but the values were significantly higher only in the case of C-ions (100%) compared to Ti and control treatments (Figure 1).

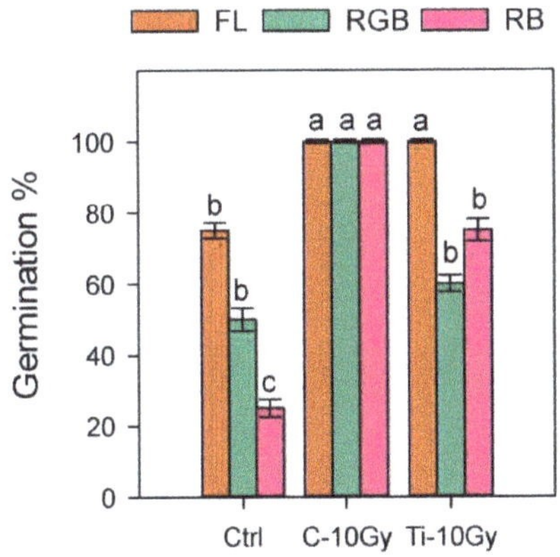

Figure 1. Percentage germination (GP) of Control (Ctrl) and irradiated Carbon (C-10 Gy) and Titanium (Ti-10 Gy) seeds of *B. vulgaris* under white fluorescent (FL), red–green–blue (RGB) and red–blue (RB) light quality regimes (n = 50). Different letters indicate statistically significant differences among treatments according to one-way ANOVA ($p < 0.05$).

IR and LQ as the main factors did not significantly affect total biomass (TB) and shoot biomass (SB) (Table 1). Contrarily, the interaction IR × LQ was significant. More specifically, the RB regime induced an increase in TB and SB ($p < 0.05$) in plants from irradiated seeds (C 10-RB, Ti 10-RB) compared to the control (Ctrl-RB).

2.2. Gas Exchanges and Chlorophyll Fluorescence Emission Measurements

The IR and LQ regimes strongly affected the photosynthetic performance of *B. vulgaris* plants as single factors and in combination. A_N and g_S of C-10 and Ti-10 plants declined ($p < 0.01$) and NPQ were raised ($p < 0.05$), while iWUE, ϕ_{PSII}, and F_v/F_m did not change significantly compared to control (Table 2). LQ, as a single factor, reduced A_N ($p < 0.05$), g_S, and ϕ_{PSII}, ($p < 0.01$) in RGB and RB compared to FL plants and an increased iWUE($p < 0.05$) and NPQ ($p < 0.001$) (Table 2). No variation was observed in F_v/F_m regardless of the IR treatments and LQ regimes (Table 2, Figure 2f).

The interaction IR × LQ (under FL regime) showed a significant decline of A_N ($p < 0.001$), g_S ($p < 0.001$), and iWUE ($p < 0.05$) in both C-10- and Ti-10-irradiated plants compared to control (Figure 2a–c).

Growth under RGB and RB regimes did not induce any differences in A_N and iWUE among irradiated plants and respective controls (Figure 2a–c), while g_S showed the lowest ($p < 0.01$) value in C10-RGB plants (Figure 2b).

Within the C-10 plant group, RGB and RB regimes reduced ($p < 0.001$) A_N and g_S compared to FL but did not influence iWUE (Figure 2a–c). In C-10 plants, A_N was unaffected by LQ, while RGB and RB regimes induced a decline ($p < 0.001$) in g_S and an increase ($p < 0.05$) of iWUE compared to FL (Figure 2a–c). Within the Ti-10 plant group, the RGB and RB light regimes enhanced A_N ($p < 0.05$), g_S ($p < 0.05$), and iWUE ($p < 0.01$) compared to FL (Figure 2a–c).

Table 1. Analysis of variance and comparison of means for total biomass (TB) and shoot biomass (SB) of *B. vulgaris* plants sprouted from Control (Ctrl) and irradiated Carbon (C-10 Gy) and Titanium (Ti-10 Gy) seeds, and grown under white fluorescent (FL), red–green–blue (RGB) and red–blue (RB) light quality regimes.

	TB	SB
IR		
Ctrl	25 ^{a}	21 ^{a}
C-10	27 ^{a}	21 ^{a}
Ti-10	27 ^{a}	21 ^{a}
LQ		
FL	29 ^{a}	23 ^{a}
RGB	24 ^{a}	19 ^{a}
RB	26 ^{a}	21 ^{a}
IR × LQ		
Ctrl-FL	31 ^{a}	25 ^{a}
Ctrl-RGB	28 ^{a}	24 ^{a}
Ctrl-RB	17 ^{b}	14 ^{b}
C 10-FL	28 ^{a}	22 ^{a}
C 10-RGB	25 ^{a}	21 ^{a}
C 10-RB	26 ^{a}	21 ^{a}
Ti 10-FL	30 ^{a}	22 ^{a}
Ti 10-RGB	23 ^{a}	19 ^{a}
Ti 10-RB	29 ^{a}	22 ^{a}
Significance		
IR	NS	NS
LQ	NS	NS
IR × LQ	*	*

TB—Total biomass, g FW plant^{-1}; SB—Shoot biomass, g FW plant^{-1}. Different letters in each column indicate significant differences according to Duncan's test ($p < 0.05$). NS—not significant; * $p < 0.05$.

Table 2. Analysis of variance and comparison of means for net CO_2 assimilation (A_N), stomatal conductance to water (g_S), intrinsic water use efficiency (iWUE), quantum yield of PSII electron transport, ϕ_{PSII}, non-photochemical quenching (NPQ), maximum PSII photochemical efficiency, (F_v/F_m) of *B. vulgaris* plants sprouted from Control (Ctrl) and irradiated Carbon (C-10 Gy) and Titanium (Ti-10 Gy) seeds, and grown under white fluorescent (FL), red–green–blue (RGB) and red–blue (RB) light quality regimes.

	A_N	g_s	iWUE	ϕ_{PSII}	NPQ	F_v/F_m
IR						
Ctrl	9.1 ^{a}	0.16 ^{a}	59 ^{a}	0.34 ^{a}	2.4 ^{b}	0.76 ^{a}
C-10	6.7 ^{b}	0.13 ^{b}	53 ^{a}	0.34 ^{a}	2.7 ^{a}	0.75 ^{a}
Ti-10	6.4 ^{b}	0.13 ^{b}	56 ^{a}	0.29 ^{a}	2.8 ^{a}	0.74 ^{a}
LQ						
FL	7.9 ^{a}	0.16 ^{a}	48 ^{b}	0.38 ^{a}	2.2 ^{b}	0.75 ^{a}
RGB	7.4 ^{b}	0.14 ^{b}	55 ^{b}	0.31 ^{b}	2.8 ^{a}	0.75 ^{a}
RB	6.8 ^{c}	0.12 ^{c}	65 ^{a}	0.28 ^{b}	2.9 ^{a}	0.75 ^{a}
Significance						
IR	***	***	NS	NS	**	NS
LQ	*	***	**	*	***	NS
IR × LQ	***	***	NS	NS	***	NS

A_N—net CO_2 assimilation (μmol CO_2 m^{-2} s^{-1}); g_S—stomatal conductance to water (mol H_2O m^{-2}s^{-1}), iWUE—intrinsic water use efficiency (μmol CO_2 mol^{-1} H_2O); ϕ_{PSII}—quantum yield of PSII electron transport; NPQ—non-photochemical quenching; F_v/F_m—maximum PSII photochemical efficiency. Different letters in the columns indicate significant differences according to Duncan's test ($p < 0.05$). NS—not significant; * $p < 0.05$; ** $p < 0.01$; *** $p < 0.001$.

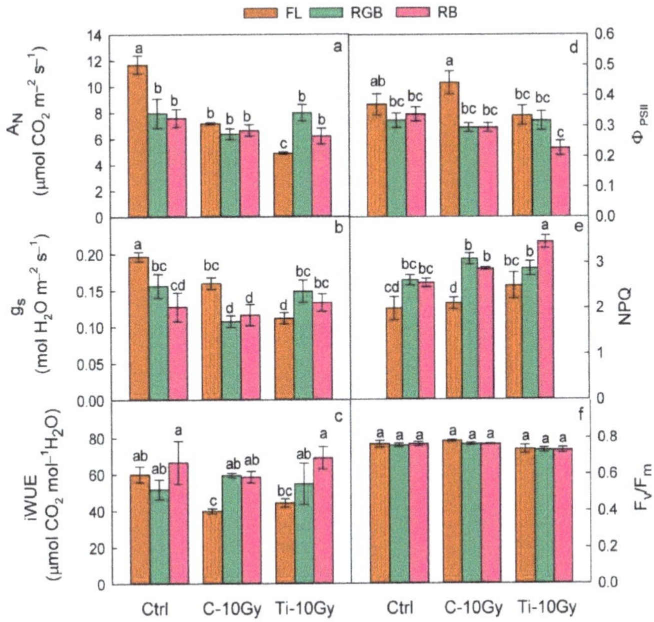

Figure 2. (**a**) Net CO_2 assimilation, A_N; (**b**) stomatal conductance, g_S; (**c**) intrinsic water use efficiency, iWUE; (**d**) quantum yield of PSII electron transport, ϕ_{PSII}; (**e**) non-photochemical quenching, NPQ; (**f**) maximum PSII photochemical efficiency, F_v/F_m of *B. vulgaris* plants sprouted from Control (Ctrl) and irradiated Carbon (C-10 Gy), and Titanium (Ti-10 Gy) seeds and grown under white fluorescent (FL), red–green–blue (RGB) and red-blue (RB) light quality regimes ($n = 5$). Different letters indicate statistically significant differences among light treatments according to one-way ANOVA ($p < 0.05$).

In all LQ regimes, ϕ_{PSII} and F_v/F_m ratio were not significantly affected by ionizing radiation. Contrary, among RB plants, Ti-ions significantly increased NPQ (Figure 2d–f). Within control plants, LQ regimes did not affect chard photochemistry. On the contrary, within the C-10 and Ti-10 plant groups, the RGB and RB regimes reduced ($p < 0.01$) ϕ_{PSII} and increased ($p < 0.01$) NPQ compared to FL. The strongest reduction of ϕ_{PSII} and the highest rise of NPQ were measured in Ti-10 RB plants (Figure 2d–f).

2.3. Plants Nutritional Traits and Bioactive Compounds

IR treatments and LQ regimes, as a single factor or interaction, determined a substantial variation in the concentration of photosynthetic pigment, total carbohydrate, proteins, and antioxidants (Table 3).

IR induced an increase ($p < 0.001$) in CHL and CAR concentration and a reduction ($p < 0.01$) in POL and ANTH in Ti-10 compared to Ctrl and C-10 plants. On the contrary, compared to control, C-10 plants showed comparable concentration of CHL, CAR, POL, and ANTH, lower ($p < 0.05$) CARB and PROT content, and higher ($p < 0.001$) level of TAC.

LQ as a single factor determined a reduction ($p < 0.001$) in CHL, CAR, and CARB under RGB and RB regimes ($p < 0.001$) compared to FL (Table 3), while for PROT and POL, the RGB plant group showed the lowest value. No significant difference was detected between FL and RB regimes. However, for ANTH and TAC, the different LQ regimes exerted diverse responses: RGB reduced ($p < 0.01$) ANTH content compared to FL and RB but increased TAC ($p < 0.01$).

Table 3. Analysis of variance and comparison of means for chlorophylls (CHL), carotenoids (CAR), carbohydrates (CARB), proteins (PROT), polyphenols (POL), anthocyanins (ANTH), and total antioxidant capacity (TAC) of *B. vulgaris* plants sprouted from Control (Ctrl) and irradiated Carbon (C-10 Gy) and Titanium (Ti-10 Gy) seeds, and grown under white fluorescent (FL), red–green–blue (RGB) and red–blue (RB) light quality regimes (n = 5).

	CHL	CAR	CARB	PROT	POL	ANTH	TAC
IR							
Ctrl	38 ^{b}	5.7 ^{b}	36 ^{a}	480 ^{a}	0.91 ^{a}	2.8 ^{a}	2.8 ^{c}
C-10	41 ^{b}	6.7 ^{b}	33 ^{b}	347 ^{b}	0.94 ^{a}	2.9 ^{a}	4.1 ^{a}
Ti-10	47 ^{a}	8.4 ^{a}	31 ^{b}	321 ^{b}	0.47 ^{b}	2.0 ^{b}	3.3 ^{b}
LQ							
FL	54 ^{a}	8.8 ^{a}	42 ^{a}	427 ^{a}	0.81 ^{a}	2.5 ^{b}	3.0 ^{b}
RGB	37 ^{b}	5.7 ^{b}	33 ^{b}	337 ^{b}	0.70 ^{b}	1.9 ^{c}	4.3 ^{a}
RB	36 ^{b}	5.9 ^{b}	26 c	381 ^{a}	0.80 ^{a}	3.3 ^{a}	2.9 ^{b}
Significance							
IR	***	***	*	***	***	***	***
LQ	***	***	***	*	*	***	***
IR × LQ	***	***	***	**	**	NS	***

CHL—Chlorophylls (µg cm^{-2}); CAR—Carotenoids (µg cm^{-2}); CARB—Carbohydrates (mg GLU eq g^{-1} FW); PROT—Proteins (µg BSA eq g^{-1} FW); POL—Polyphenols (mg GAE g^{-1} FW); ANTH—Anthocyanins ((A_{530}-1/3A_{657}) g^{-1} FW); TAC—Total antioxidant capacity (µmol Trolox eq g^{-1} FW). Different letters in each column indicate significant differences according to Duncan's test ($p < 0.05$). NS—not significant; * $p < 0.05$; ** $p < 0.01$; *** $p < 0.001$.

The analysis of interactions IR × LQ highlighted that both control and irradiated plants are characterized by lower ($p < 0.001$) CHL and CAR content under RGB and RB than under FL regime (Figure 3a,b).

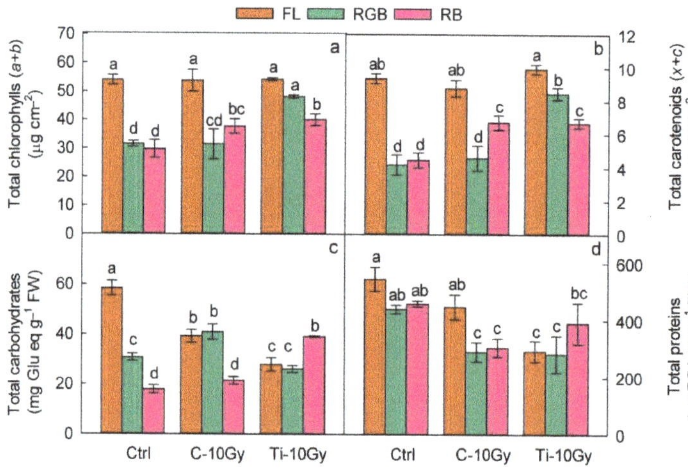

Figure 3. (**a**) Total chlorophylls; (**b**) total carotenoids; (**c**) total carbohydrates; (**d**) total proteins of *B. vulgaris* plants sprouted from Control (Ctrl) and irradiated Carbon (C-10 Gy), and Titanium (Ti-10 Gy) seeds and grown under white fluorescent (FL), red–green–blue (RGB) and red–blue (RB) light quality regimes (n = 5). Different letters indicate statistically significant differences among treatments according to one-way ANOVA ($p < 0.05$).

Total carbohydrate content was affected by IR and LQ; indeed, it decreased ($p < 0.05$) in irradiated plants under the FL regime compared to control, while under RB, in Ti-10 plants, it was higher than in control (Figure 3c). C-FL plants showed the highest ($p < 0.001$)

concentration of carbohydrates. In irradiated C-10 plants, the highest increase ($p < 0.001$) of carbohydrates was obtained under RGB regime, whereas within the Ti-10 plant group, the highest ($p < 0.001$) carbohydrate concentration was determined under the RB regime (Figure 3c).

The interaction IR × LQ consistently affected the protein content. Within the control and Ti-10 plant group, the protein content was not affected by LQ regimes. On the contrary, in the C-10 Gy plant group, the protein amount declined ($p < 0.05$) under RGB and RB compared to FL (Figure 3d).

The comparison among control and irradiated plants at the same LQ regimes showed that the RGB reduced ($p < 0.05$) the total protein content in both irradiated plants compared to control. Otherwise, under FL and RB, the total protein concentration decreased only in Ti-10-FL ($p < 0.05$) and C-10-RB ($p < 0.05$) plants, respectively (Figure 3d).

The interaction IR × LQ indicated that polyphenols significantly declined ($p < 0.001$) in T-10 plants regardless of LQ quality (Figure 4a). Under the FL regime, anthocyanins decreased ($p < 0.01$) in T-10-FL plants compared to Ctrl-FL plants. Conversely, under RGB and RB regimes, no significant difference in anthocyanin level was found between control and irradiated plants (Figure 4b). In the Ctrl plant group, anthocyanin concentration was lower ($p < 0.05$) under RGB than under FL and RB, whereas in C-10 and Ti-10 irradiated groups, anthocyanin levels increased under RB compared to FL and RGB regimes (Figure 4b).

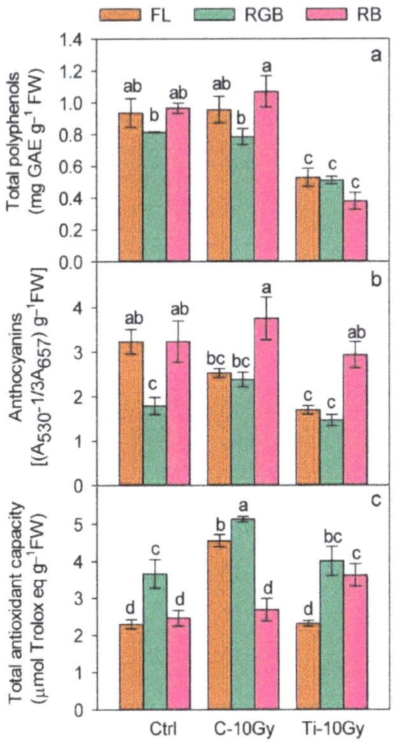

Figure 4. (**a**) Total polyphenol; (**b**) anthocyanin content; (**c**) total antioxidant capacity of *B. vulgaris* plants sprouted from Control (C) and irradiated Carbon (C-10 Gy), and Titanium (Ti-10 Gy) seeds and grown under white fluorescent (FL), red–green–blue (RGB) and red–blue (RB) light quality regimes ($n = 5$). Different letters indicate statistically significant differences among treatments according to one-way ANOVA ($p < 0.05$).

Finally, the total antioxidant capacity within the control group increased ($p < 0.05$) under the RGB compared to FL and RB plants; in the C-10 plant group, the highest ($p < 0.01$) value was measured under the RGB regime, while in the Ti-10 plant group, the highest values were found under both RGB and FL (Figure 4c). The comparison among control and irradiated plants under the same LQ regime evidenced that under FL and RGB, C-10 plants showed the highest values of TAC ($p < 0.001$, $p < 0.01$, respectively). On the other hand, the RB regime promoted ($p < 0.01$) TAC in Ti-10 plants (Figure 4c).

2.4. Heatmap Analysis

An overview of all measured parameters in response to heavy ions irradiation (C-10 and Ti-10) and the three LQ regimes (FL, RGB, and RB) is reported in Figure 5.

The heatmap identified two main clusters. The first cluster (I) included plants sprouted from control and C-10-irradiated seeds grown under the FL regime. The second cluster (II) was split into two sub-clusters: the first incorporated Ti-10 irradiated plants grown under FL and RGB regimes; the second included control and C-10 plants grown under RB, control, and C-10 plants grown under RGB and Ti-10 plants grown under RB. The heatmap indicated that control and C-10 irradiated plants showed a similar response for different physiological and biochemical attributes regardless of the LQ regime. In particular, under FL, plants exhibited the highest values of biomass, pigments, PSII photochemical efficiency, and antioxidant compounds. Within the Ti-10 group, RB plants were characterized by higher values of iWUE and NPQ than FL and RGB plants. Finally, the heatmap indicated that LQ regimes exerted different physiological responses in *B. vulgaris* plants irradiated with C- or Ti-ions.

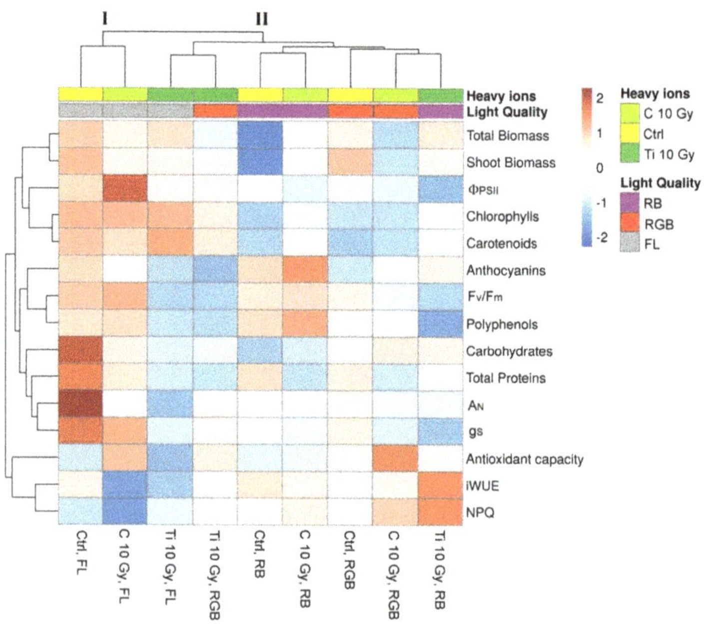

Figure 5. Cluster heatmap analysis summarizing morphological, eco-physiological, and biochemical parameters of *Beta vulgaris* L. cv cicla plants sprouted from Control (Ctrl) and irradiated Carbon (C-10 Gy), and Titanium (Ti-10 Gy) seeds and grown under white fluorescent (FL), red-green-blue (RGB) and red-blue (RB) LQ regimes. Numeric differences within the data matrix are shownby the color scale: light blue and dark blue indicate increasing and decreasing values, respectively. Parameters are clustered in the rows; sample groups are clustered in the columns by the two independent factors: IR treatment and LQ regimes.

3. Discussion

This work highlighted that a low dose of carbon and titanium heavy ions, delivered at the seed stage, may modify the *B. vulgaris* eco-physiological response (i.e., photosynthesis and accumulation of bioactive compounds) depending on different light quality regimes during growth. These results may have implications on controlled environment agriculture, especially in extreme environments such as Space.

In the view of space cultivation, seed germination could represent a critical step. Previous research demonstrated that seed responses to IR depend on the plant species and the type of ion. The seed irradiation with C-ions at the dose of 10 Gy, reduced the germination rate in rice and bean [6,47], whereas no variation occurred in spinach for doses up to 15 Gy [48]. Conversely, there is still little information about the effect of Ti-ions on different species. Ti-ions did not affect the germination of bean seeds at the dose of 10 Gy [6]. In *B. vulgaris* seeds, the irradiation with C-ions promoted the percentage germination (100%) compared to control and Ti-ions, regardless of the LQ regimes. These results indicated that more energetic C-ions likely favored seed tegument porosity which, in turn, may have promoted germination through a higher water permeability [4,49]. In control plants, the reduction of germination under RGB and RB light regimes could be ascribed to the higher incidence of blue wavelengths, which generally inactivate the phytochrome A involved in seed germination [50]. Moreover, the higher percentage germination found in Ctrl-RGB compared to Ctrl-RB seeds may be ascribed to the presence of green wavelength, which is known to promote germination through phytochrome [51].

Generally, the exposure of plants to IR determines a reduction of plant growth and biomass, inducing a more compact plant architecture [4,52–55]. On the other hand, the RB growth regime may produce an enhancement in biomass depending on the intrinsic characteristics of the species [31,56]. In our case, IR and LQ as single factors did not affect plant biomass, but their interaction produced significant differences under RB light regime. In particular, C and Ti-RB plants were characterized by higher edible biomass than control, representing a suitable trait for plants destined to grow in BLSSs.

Heavy ions and LQ regimes and their interaction deeply affected the photosynthetic activity in *B. vulgaris*. IR generally impaired A_N, g_S, and iWUE, regardless of the type of radiation and dose [52,57–60]. Moreover, these parameters are strictly interconnected, as CO_2 uptake and water use follow the same route through stomata [61]. In control plants, leaf gas exchanges were sensitive to LQ, especially under RGB and RB regimes that strongly reduced A_N and g_S compared to FL. The seed irradiation with C-ion seemed to offset the effect LQ on A_N, which resulted comparable under all regimes. It is hypothesized that in C-ion irradiated plants, the high intensity of red and blue wavelengths of RGB and RB regimes may have improved the stomatal control, ultimately resulting in the enhancement of iWUE. However, the occurrence of stomatal limitations due to the potentially detrimental effects of C heavy ions on photosynthetic machinery cannot be excluded. On the other hand, the seed irradiation with Ti-ion under RGB and RB regimes stimulated A_N and g_S compared to FL, determining, also in this case, an improvement of iWUE. In response to different LQ regimes during growth, leaves of irradiated plants have probably adopted adjustments in mesophyll traits and stomatal movements to improve photosynthesis and iWUE [45]. According to other authors, RB wavelengths, alone or supplemented with the green light that penetrates deeper inside the canopy, may have induced changes in leaf thickness, promoting the CO_2 diffusion within chloroplasts [56,62–66]. In addition, the blue wavelengths, acting on the guard cells, stimulated the stomatal opening, improving leaf conductance and consequently, photosynthesis [26,67–69].

Both IR and LQ affected not only the dark but also the light phase of photosynthesis and, more specifically, the partitioning of light energy. In RGB and RB plants, the reduction of ϕ_{PSII} was consistent with the decline of A_N and the rise of NPQ, indicating that the photosynthetic apparatus diverted the light energy in thermal dissipation mechanisms in conditions of reduced carbon assimilation [70]. On the contrary, in C-10-FL plants, the A_N decline was accompanied by high values of ϕ_{PSII}. In this case, photochemical processes different from photosynthesis (i.e., photorespiration, Mehler reaction) occurred to avoid photoinhibition and photooxidative damage to photosystems. This response suggested that under C-irradiation, plants adopted a mechanism to optimize the PSII efficiency by transferring the excess light energy, which is potentially detrimental for photosystems to the other photochemical processes [71]. Control and Ti-ion-irradiated plants showed similar photochemical behavior. However, the higher NPQ measured in Ti-10 RB plants suggested that thermal dissipation processes were amplified under Ti irradiation and used as a safety valve against putative photoinhibition damages [72,73]. The absence of difference in the maximum quantum efficiency of PSII (F_v/F_m) among all treatments confirmed the efficiency of the different regulatory mechanisms of absorbed light induced by different heavy ion treatments.

The photochemical reactions observed in RGB- and RB-irradiated plants were consistent with the lower photosynthetic pigment content. The down-regulation of chlorophyll and carotenoid biosynthesis represents a safety strategy to avoid excessive light harvesting. Since the pigment reduction occurred in both control and irradiated plants, it may be argued that it depended on LQ more than IR. Indeed, red wavelengths, being photosynthetically more efficient, usually determined a photosynthetic pigment reduction in different species [31,66,73–76].

Our study pointed out that the FL-irradiated plants exhibited a reduced carbohydrate and protein production compared to the control consistent with the decline of photosynthesis. Previous studies performed on different plant species exposed to gamma rays demonstrated that depending on dose and plant phenological stage, the carbohydrate and protein levels may decrease, remain unchanged, or increase [13,52,77–79]. Generally, the higher dosage of gamma irradiation breaks down the seed proteins releasing more amino acids. This process may inhibit protein synthesis, thus inducing a decline in plant total protein content [78,80]. Besides IR, LQ can also modulate sugar and protein production. For instance, RB may induce a reduction in sugars [81] or an enhancement of the sugar and protein content in several species [31,82,83]. The intrinsic characteristics of the specific heavy ions may have produced a different behavior under the RB regime, which exerted a positive effect only when applied to Ti-ion-irradiated plants.

IR strongly influenced TPC, ANTH, and TAC, depending on ion type. While C-ions did not affect the concentration of anthocyanins and polyphenols compared to controls, Ti-ions reduced these compounds. To counteract the IR-induced oxidative stress and mitigate the risk of disease, a diet rich in polyphenols is essential for astronauts. Usually, phenolic compounds exert a screening function against high levels of solar radiation; protecting cell structures from photoinhibition damages [37,54,84,85] the same way, they offset the detrimental effects of IR [54,75,85–87]. The effects of IR on polyphenols and anthocyanins are controversial because some irradiated plants showed an enhancement after the exposure to gamma and X-rays or C-heavy ions [54,59,75,86], while other species exhibited a decline [54,85]. The different outcomes depend on the radiation quality and dose. In our study, Ti-ions induced a decline in polyphenols content regardless of the growth LQ regime.

The anthocyanin concentration was not only affected by IR but also by LQ.

In both C- and Ti-irradiated plants, the RB regime stimulated the anthocyanin synthesis compared to FL and RGB. The biosynthesis of anthocyanins is typically associated with blue light, but it may also be stimulated by red and green wavelengths [37,88]. It may be supposed that the higher intensity of red wavelengths in RB compared to the other regimes may have boosted the anthocyanin production, improving the nutritional properties of chard as observed for many crops [89–91]. Hence, the increase in anthocyanins may be considered a desirable feature for irradiated *B. vulgaris* plants.

Finally, C-ions irradiation determined a consistent rise of the total antioxidant capacity, which may be ascribed to the production of several different compounds characterized by antioxidant properties, as for other species, such as lettuce, irradiated with UV and gamma rays [75,92,93]. The antioxidant response of Ti-ion-irradiated plants was enhanced under

RGB and RB regimes, confirming that red and blue supplemented with green wavelengths can promote the synthesis of free radical scavenger molecules. These compounds potentially improve the antioxidant capacity of chard plants and enhance their tolerance to stress conditions and nutrient quality [39,91,94,95]. Moreover, the high antioxidant content may be considered an added value, also for the use of chard as a supplement for astronauts' diet.

4. Materials and Methods

4.1. Plant Material, Irradiation Procedure, and Experimental Design

Beta vulgaris L. var. *cicla* (white chard) is a widely cultivated crop, considered a functional food because of its high content of secondary metabolites, associated with some beneficial effects, including anti-tumoral activity [96]. Moreover, its compactness and high ratio of edible biomass/wastes make chard one of the candidates' crops to be cultivated in the Space Greenhouses, designed as Closed Ecological Life Support Systems [3].

Dry chard seeds ($n = 150$) were divided into not-irradiated control ($n = 50$) and treated groups ($n = 100$). More specifically, 50 seeds were irradiated with carbonium (isotope ^{12}C; E: 300 MeV/u (monoenergetic), LET: 13 keV/µm; dose rate 1 Gy/min), and 50 seeds with titanium (isotope ^{50}Ti; E: 1000 MeV/u (monoenergetic), LET: 108 keV/µm; dose rate 1 Gy/min) at the dose of 10 Gy. In particular, C-ions are considered reference radiation and the dose of 10 Gy, being below the threshold for DNA damage, may not be lethal for plant development [19,97]. Seeds were collected into T25-flasks and the irradiation was performed using a pencil beam in a spread-out Bragg peak (SOBP), at the heavy-ion synchrotron (SIS) at the GSI Helmholtz zentrumfür Schwerionenforschung GmbH, (Darmstadt, Germany).

The seeds were maintained in the same storage and transport facilities to avoid bias due to different conditions during traveling. Irradiated and not-irradiated (control) seeds were then transferred to the laboratory and placed in Petri dishes on three layers of filter paper to follow the germination process. Both germination and plant cultivation took place in a climatized room under three different light regimens: (1) FL, provided by white fluorescent tubes (Lumilux L360W/640 and L360W/830, Osram, Germany); (2) RGB (red–green–blue) and (3) RB (red–blue) provided by red, green, and blue LEDs (Octa Light LTD, Bulgaria) with the following emission peaks: 630 nm red, 510 nm green, 440 nm blue. The spectral composition of LQ treatments (Figure 6a,b) was measured by a SR-3000A spectroradiometer at 10 nm resolution (Macam Photometrics Ltd., Livingston, Scotland, UK) at the top of the canopy.

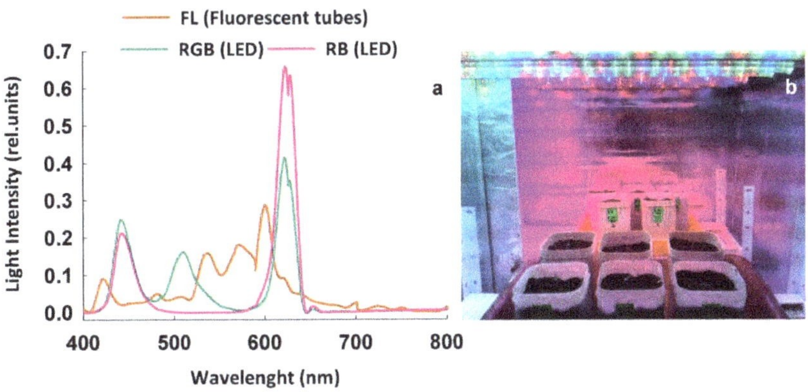

Figure 6. (**a**) Spectral distributions in the relative energy of the light quality regimes used in the study: FL—white fluorescent tubes; LED RGB—red–green–blue; LED RB—red–blue. Photon flux density: 300 µmol m^{-2}s^{-1}; (**b**) Particular of growth chambers with RGB (front) and RB (back) light quality regimes.

The total photosynthetic photon flux density (PPFD) was fixed at 300 ± 5 μmol photons $m^{-2}s^{-1}$ at the canopy level for all LQ regimes. All plants were kept under air temperature of 25/20 °C (day/night), relative humidity 60–70%, and a photoperiod of 12 h. Plants were fertilized with tap water and Hoagland's solution every two weeks.

The plant growth was followed up to 60 days after sowing (DAS) at the Plant Physiology and Genetics Institute of the Bulgarian Academy of Science (Sofia, Bulgaria). Gas exchanges and fluorescence emission measurements were carried out 60 days after sowing (DAS) on fully expanded leaves to assess how radiation may have affected the functionality of the photosynthetic apparatus and if the plant growth under different LQ may have influenced the plant photosynthetic behavior. In addition, at the end of the vegetative cycle (60 DAS), biometrical measurements, leaf functional traits, photosynthetic pigments, total carbohydrates, and antioxidant content were also determined on mature leaves as a proxy for carbon allocation and plant nutritional status. These analyses were performed at the Department of Biology of University Federico II of Naples (Italy).

Figure 7 displays the schema of the experimental design of the study.

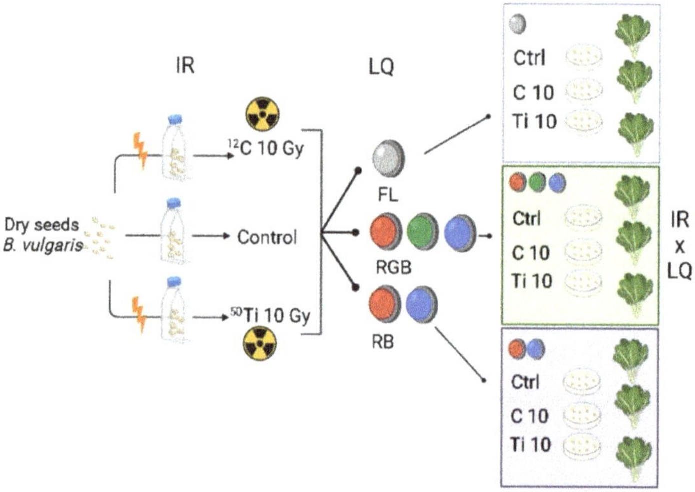

Figure 7. Schema of the experimental design created with BioRender.com (accessed on 2 May 2022).

4.2. Germination and Biometrical Measurements

The percentage of seed germination under FL, RGB, and RBLQ regimes was evaluated. Seeds were considered germinated when the root protruded from the seed coat. The percentage germination (GP) was calculated at 7 days after sowing (DAS), according to the formula:

$$GP_{7DAS} = [(\text{Number of germinated seeds after 7 days})/\text{Total seed number}] \times 100 \quad (1)$$

After germination, 10 control, 10 C—ion, and 10 Ti—ion—irradiated seedlings were sown in 3.0 L pots filled with peat soil. At the end of the experimental period, 60 DAS, total biomass (TB), and shoot biomass (SB) were determined on five plants for each treatment, weighing the whole plant and the shoot portion, respectively. The TB and SB were expressed as grams of fresh weight per plant (g FW $plant^{-1}$).

4.3. Leaf Gas Exchange and Chlorophyll aFluorescence Emission Measurements

Leaf gas exchange was measured at 60 DAS on five fully expanded leaves from five plants per treatment (one leaf per plant) by a portable gas-exchange system (LCpro+, ADC BioScientific, UK). The middle part of the leaf was clamped into the 6.25 cm^2 gas-exchange cuvette and exposed to a constant flow of 300 μmol s^{-1} of synthetic air (79% N_2, 21% O_2, and 400 μmol mol^{-1} CO_2). Measurements were carried out at 25 ± 2 °C leaf temperature and 500 μmol $m^{-2}s^{-1}$ photosynthetic photon flux density (PPFD). The relative humidity in the leaf chamber was set at 50–60%. The intrinsic water use efficiency (iWUE) was calculated as a ratio between photosynthesis (A_N) and stomatal conductance to water (g_s). All gas-exchange parameters were recorded after reaching a steady-state, usually 7–10 min for each measurement, and calculated by the equations of von Caemmerer and Farquhar [98] with the software operating within the gas-exchange system.

Chlorophyll *a* fluorescence measurements were carried out on five fully expanded leaves from five plants per treatment (one leaf per plant) using a Fluorescence Monitoring System (FMS, Hansathech Instruments, King's Lynn, UK). The determination of minimum (F_o) and maximum (F_m) fluorescence was carried out on 30 min dark adapted leaves. The maximum quantum yield of PSII photochemistry (F_v/F_m) was determined as $(F_m-F_o)/F_m$. The measurements in the light were carried out on leaves adapted to PPFD of 500 μmol m^{-2} s^{-1}. A saturating pulse of 0.8 s with >6000 μmol photons m^{-2} s^{-1} was applied to determine the maximum (F_m') and the steady-state (F_s) fluorescence in light adapted conditions. The quantum yield of PSII electron transport (ϕ_{PSII}) was calculated according to Genty et al. [99] as: $\phi_{PSII} = (F_m' - F_s)/F_m'$. The non-photochemical quenching (NPQ) was calculated as $NPQ = (F_m - F_m')/F_m'$ [100].

4.4. Determinationof Photosynthetic Pigments and Antioxidants

Photosynthetic pigment and antioxidant contents were determined on five fresh leaves collected from five different plants (one leaf per plant) from each experimental condition.

The determination of the photosynthetic pigments content, namely total chlorophylls (*a* + *b*) and carotenoids (*x* + *c*), was performed according to Lichtenthaler [101]. Leaf samples of the known area (0.283 cm^2) were homogenized in ice-cold 100% acetone using a mortar and pestle. The extracts were centrifuged at 5000 rpm for 5 min in a Labofuge GL (Heraeus Sepatech, Hanau, Germany). The sample absorbance was measured by a spectrophotometer (UV-VIS Cary 100; Agilent Technologies) at wavelengths of 470, 645, and 662 nm. Pigment concentration was expressed as μg cm^{-2}.

The total polyphenol content was evaluated on fresh samples powdered in liquid nitrogen. Samples were extracted in 80% methanol at 4 °C, centrifuged at 11,000 rpm for 5 min. The soluble fraction was mixed with 10% Folin–Ciocâlteu solution, 1:1 *v*/*v*, and after 3 min, 700 mM Na_2CO_3 solution was added to the resulting mixture (1:5, *v*/*v*). Samples were incubated for 2 h in the darkness. The absorbance was measured at 765 nm with a spectrophotometer (UV-VIS Cary 100; Agilent Technologies). The total polyphenol amount was expressed as mg of Gallic Acid Equivalents g^{-1} FW (mg GAE g^{-1} FW) using a gallic acid standard curve.

The anthocyanin content was determined on fresh samples powdered in liquid nitrogen, treated with methanol 1% HCl solution and stored overnight at 4 °C. After adding 1:0.6 (*v*/*v*) ultra-pure water and chloroform at 1:1.6, *v*/*v*), samples were centrifuged at 11,000 rpm for 5 min. After mixing the supernatant with 1:1 (*v*/*v*) (60% (methanol 1% HCl) 40% ultra-pure water)solution, the absorbance was measured using a spectrophotometer (UV-VIS Cary 100, Agilent Technologies, Palo Alto, CA, USA) at 530 and 657 nm. The relative amount of anthocyanin was expressed as (A_{530}—1/3A_{657}) g^{-1} FW [102].

The antioxidant capacity was assessed by the Ferric Reducing Antioxidant Power (FRAP) assay, performed on fresh leaves powdered in liquid nitrogen, according to George et al. [103]. Briefly, the samples (0.250 g) were treated with methanol/water solution (60:40, *v*/*v*) and centrifuged at 14,000 rpm for 15 min at 4 °C. The extracts were mixed with the FRAP reagents (300 mM acetate buffer pH 3.6, 1:16.6 *v*/*v*; 10 mM tripyridyltriazine, TPTZ, in 40 mM HCl, 1:1.6 *v*/*v*; 12 mM $FeCl_3$, 1:1.6 *v*/*v*) and incubated for 1 h in the dark. Then, the absorbance was read at 593 nm by a spectrophotometer (UV-VIS Cary 100; Agilent Technologies). The antioxidant capacity was calculated using a Trolox standard curve and expressed as μmol Trolox equivalents g^{-1} FW (μmol Trolox eq g^{-1} FW).

4.5. Total Soluble Carbohydrate Content and Protein Quantification

Total soluble carbohydrates were determined on five leaves for each treatment by the anthrone method, as reported in Hedge and Hofreiter [104]. The absorbance was measured at 630 nm using a spectrophotometer (UV-VIS Cary 100, Agilent Technologies, Palo Alto, CA, USA). The amount of soluble carbohydrates in the extracts was expressed as mg Glucose equivalents g^{-1} FW (mg Glu eq g^{-1} FW) using a Glucose standard curve.

Protein extraction was carried out on five fresh leaf samples ground in liquid nitrogen, according to Wang et al. [105]. Total protein content was quantified by Bradford colorimetric assay [106], measuring the sample absorbance at 595 nm by a spectrophotometer (UV-VIS Cary 100, Agilent Technologies, Palo Alto, CA, USA). Using a BSA standard curve, the protein concentration was expressed as μg BSA (bovine serum albumin) equivalents g^{-1} FW.

4.6. Statistical Analysis

Statistical data analysis was performed using SigmaPlot 12 software (Jandel Scientific, San Rafael, CA, USA). The effect of IR (C- and Ti-ions) and LQ regimes (FL, RGB, RB) on morphological, ecophysiological, and biochemical traits of chard plants were evaluated by processing data by two-way analysis of variance (ANOVA) followed by Duncan multiple comparison tests ($p < 0.05$). The Kolmogorov–Smirnov test was used to check the normality. When the interaction between the two factors (IR × LQ) was significant, data were further processed by applying one-way ANOVA, and multiple comparison tests were done with the Duncan test.

The overall parameters were visualized by a heatmap, plotted using the ClustVis program package (https://biit.cs.ut.ee/clustvis/online, accessed on 20 June 2022) and clustering both rows and columns with Euclidean distance and average linkage. In the heatmap, the numeric differences were evidenced by a color scale: red and blue indicate increasing and decreasing values, respectively.

5. Conclusions

The irradiation of dry chard seeds with carbon or titanium high-energy ions significantly modified the plant response to light quality. In particular, under the FL regime, gas exchanges of C- and Ti-ion-irradiated plants strongly declined compared to control. However, control and C-ion-irradiated plants showed a physiological performance higher than titanium plants in terms of for pigments content, PSII photochemical efficiency, and bioactive compounds.

The growth under RGB and RB regimes offset the differences of gas exchanges between control and C- and Ti-ion plants. C-ions induced the strongest antioxidant response regardless of light quality regimes. Furthermore, the interaction Ti-ion × RB was effective in improving iWUE, and the production of pigments, carbohydrates, and antioxidants.

The overall results indicate that by manipulating the interaction IR × LQ, it is possible to regulate the photosynthesis in order to obtain plants that are more performing in resource regeneration linked to gas exchanges (CO_2 removal, O_2 production) but also to modify the bioactive compound amounts in leaf edible tissues, which may result in a beneficial outcome for the astronauts' diet.

Author Contributions: Conceptualization, C.A. (Carmen Arena) and V.D.M.; methodology, C.A. (Carmen Arena), V.V., T.T., E.V., C.A. (Chiara Amitrano), L.G.I. and P.S.; formal analysis, T.T. and E.V.; investigation, V.V., T.T., L.G.I., E.V., C.A. (Carmen Arena), P.S. and C.A. (Chiara Amitrano); resources, V.V. and C.A. (Carmen Arena); data curation, E.V., T.T. and C.A. (Carmen Arena); writing—original draft preparation, E.V. and C.A. (Carmen Arena); writing—review and editing, all the authors; supervision, C.A. (Carmen Arena), V.V. and V.D.M. All authors have read and agreed to the published version of the manuscript.

Funding: This research received no external funding.

Institutional Review Board Statement: Not applicable.

Informed Consent Statement: Not applicable.

Data Availability Statement: The data supporting the results of this study are accessible from the corresponding author (C.A. (Carmen Arena)) upon reasonable request.

Acknowledgments: The results presented here are based on the experiment SBio08_DeMicco performed at SIS18/Cave A at the GSI Helmholtz zentrum für Schwerionenforschung, Darmstadt (Germany) in the frame of FAIR Phase-0. The authors are grateful to Marco Durante, Head of the Biophysics Department at GSI Helmholtz zentrum für Schwerionenforschung for the irradiation of the samples, and for the scientific and technical support during the irradiation procedure.

Conflicts of Interest: The authors declare no conflict of interest.

References

1. Wolff, S.A.; Coelho, L.H.; Zabrodina, M.; Brinckmann, E.; Kittang, A.I. Plant mineral nutrition, gas exchanges and photosynthesis in space: A review. *Adv. Space Res.* **2013**, *51*, 465–475. [CrossRef]
2. Wheeler, R.M. Agriculture for space: People and places paving the way. *Open Agric.* **2017**, *2*, 14–32. [CrossRef]
3. Zabel, P. Designing a Closed Ecological Life Support System for Plants, Overview. In *Handbook of Life Support Systems for Spacecraft and Extraterrestrial Habits*; Seedhouse, E., Shayler, D., Eds.; Springer: Cham, Switzerland, 2018.
4. Arena, C.; Vitale, E.; Hay Mele, B.; Cataletto, P.R.; Turano, M.; Simoniello, P.; De Micco, V. Suitability of *Solanum lycopersicum* L. 'Microtom' to be grown in Bioregenerative Life Support Systems: Exploring the effect of high-LET ionising radiation on photosynthesis, leaf structure and fruit traits. *Plant Biol.* **2019**, *21*, 615–626. [CrossRef] [PubMed]
5. Arena, C.; De Micco, V.; De Maio, A. Growth alteration and leaf biochemical responses in *Phaseolus vulgaris* exposed to different doses of ionising radiation. *Plant Biol.* **2014**, *161*, 194–202. [CrossRef] [PubMed]
6. De Micco, V.; Arena, C.; Pignalosa, D.; Durante, M. Effects of sparsely and densely ionizing radiation on plants. *Radiat. Environ. Biophys.* **2011**, *50*, 1–19. [CrossRef]
7. Hwang, J.E.; Hwang, S.G.; Kim, S.E.; Lee, K.J.; Jang, C.S.; Kim, J.B.; Kim, S.H.; Ha, B.K.; Ahn, J.W.; Kang, S.Y.; et al. Transcriptome profiling in response to different types of ionizing radiation and identification of multiple radio marker genes in rice. *Physiol. Plant.* **2014**, *150*, 606–679. [CrossRef]
8. Park, Y.C.; Kim, J.J.; Kim, D.S.; Jang, C.S. Rice RING E3 ligase may negatively regulate gamma-ray response to mediate the degradation of photosynthesis-related proteins. *Planta* **2015**, *241*, 1119–1129. [CrossRef]
9. Arena, C.; Turano, M.; Hay Mele, B.; Cataletto, P.R.; Furia, M.; Pugliese, M.; De Micco, V. Anatomy, photochemical activity, and DNA polymorphism in leaves of dwarf tomato irradiated with X-rays. *Biol. Plant.* **2017**, *61*, 205–314. [CrossRef]
10. Zaka, R.; Vandecasteele, C.M.; Misset, T.M. Effect of flow chronic doses of ionizing radiation on antioxidant enzymes and G6PDH activities in Stipacapillata (Poaceae). *J. Exp. Bot.* **2002**, *53*, 1979–1987. [CrossRef]
11. Esnault, M.A.; Legue, F.; Chenal, C. Ionizing radiation: Advances in plant response. *Environ. Exp. Bot.* **2010**, *68*, 23–237. [CrossRef]
12. Kim, J.H.; Moon, Y.R.; Lee, M.H.; Kim, J.H.; Wi, S.G.; Park, B.J.; Kim, C.S.; Chung, B.Y. Photosynthetic capacity of Arabidopsis Plants at the reproductive stage tolerates γ-irradiation. *J. Radiat. Res.* **2011**, *52*, 441–449. [CrossRef] [PubMed]
13. Jan, S.; Parween, T.; Siddiqqi, T.O.; Mahmooduzzafar. Effect of gamma radiation on morphological, biochemical, and physiological aspects of plants and plant products. *Environ. Rev.* **2012**, *20*, 17–39. [CrossRef]
14. Schuy, C.; Weber, U.; Durante, M. Hybrid active-passive space radiation simulation concept for GSI and the future FAIR facility. *Front. Phys.* **2020**, *8*, 337. [CrossRef]
15. Simonsen, L.C.; Slaba, T.C.; Guida, P.; Rusek, A. NASA's first ground-based Galactic Cosmic Ray Simulator: Enabling a new era in space radiobiology research. *PLoS Biol.* **2020**, *18*, e3000669. [CrossRef]
16. Jangiam, W.; Tungjai, M.; Rithidech, K.N. Induction of chronic oxidative stress, chronic inflammation and aberrant patterns of DNA methylation in the liver of titanium exposed CBA/Caj mice. *Int. J. Radiat. Biol.* **2015**, *91*, 389–398. [CrossRef]
17. Rithidech, K.N.; Jangiam, W.; Tungjai, M.; Gordon, C.; Honikei, L.; Whorton, E.B. Induction of chronic inflammation and altered levels of DNA Hydroxymethylation in Somatic and Germinal tissue of CBA/CaJ Mice exposed to ^{48}Ti ions. *Front. Oncol.* **2016**, *6*, 155. [CrossRef]
18. Li, Z.; Jella, K.K.; Jaafar, L.; Li, S.; Park, S.; Story, M.D.; Wang, H.; Wang, Y.; Dynan, W.S. Exposure to galactic cosmic radiation compromises DNA repair and increases the potential for oncogenic chromosomal. *Sci. Rep.* **2018**, *8*, 11038. [CrossRef]
19. De Micco, V.; De Francesco, S.; Amitrano, C.; Arena, C. Comparative Analysis of the Effect of Carbon- and Titanium-Ions Irradiation on Morpho-Anatomical and Biochemical Traits of *Dolichosmelanophthalmus* DC. Seedlings Aimed to Space Exploration. *Plants* **2021**, *10*, 2272. [CrossRef]
20. Ouzounis, T.; Rosenqvist, E.; Ottesen, C.O. Spectral effects of artificial light on plant physiology and secondary metabolism: A review. *HortScience* **2015**, *50*, 1128–1135. [CrossRef]
21. Barrero, J.M.; Jacobsen, J.V.; Talbot, M.J.; White, R.G.; Swain, S.M.; Garvin, D.F.; Gubler, F. Grain dormancy and light quality effects on germination in the model grass *Brachypodium distachyon*. *New Phytol.* **2012**, *193*, 376–386. [CrossRef]
22. Liu, X.Y.; Guo, S.R.; Xu, Z.G.; Jiao, X.L.; Takafumi, T. Regulation of chloroplast ultrastructure, cross-section anatomy of leaves, and morphology of stomata of cherry tomato by light-emitting diodes. *HortScience* **2011**, *45*, 217–221.

23. Singh, D.; Basu, C.; Meinhardt-Wollweber, M.; Roth, B. LEDs for energy efficient greenhouse lighting. *Renew. Sust. Energ. Rev.* **2015**, *49*, 139–147. [CrossRef]
24. Izzo, L.G.; Mickens, M.A.; Aronne, G.; Gómez, C. Spectral effects of blue and red light on growth, anatomy, and physiology of lettuce. *Physiol. Plant.* **2021**, *172*, 2191–2202. [CrossRef]
25. Olle, M.; Viršilė, A. The effects of light-emitting diode lighting on greenhouse plant growth and quality. *Agric. Food Sci.* **2013**, *22*, 223–234. [CrossRef]
26. Goins, G.D.; Yorio, N.C.; Sanwo, M.M.; Brown, C.S. Photomorphogenesis, photosynthesis, and seed yield of wheat plants grown under red light-emitting diodes (LED) with and without supplemental blue light. *J. Exp. Bot.* **1997**, *48*, 1407–1413. [CrossRef] [PubMed]
27. Arena, C.; Tsonev, T.; Doneva, D.; De Micco, V.; Michelozzi, M.; Brunetti, C.; Centritto, M.; Fineschi, S.; Velikova, V.; Loreto, F. The effect of light quality on growth, photosynthesis, leaf anatomy and volatile isoprenoids of a monoterpene-emitting herbaceous species (*Solanum lycopersicum* L.) and an isoprene-emitting tree (*Platanusorientalis* L.). *Environ. Exp. Bot.* **2016**, *130*, 122–132. [CrossRef]
28. Trouwborst, G.; Hogewoning, S.W.; van Kooten, O.; Harbinson, J.; van Ieperen, W. Plasticity of photosynthesis after the red-light syndrome in cucumber. *Environ. Exp. Bot.* **2016**, *121*, 12175–12182. [CrossRef]
29. Izzo, L.G.; Hay Mele, B.; Vitale, L.; Vitale, E.; Arena, C. The role of monochromatic red and blue light in tomato early photomorphogenesis and photosynthetic traits. *Environ. Exp. Bot.* **2020**, *179*, 104195. [CrossRef]
30. Fan, X.X.; Xu, Z.G.; Liu, X.Y.; Tang, C.M.; Wang, L.W.; Han, V.L. Effects of light intensity on the growth and leaf development of young tomato plants grown under a combination of red and blue light. *Sci. Hortic.* **2013**, *153*, 50–55. [CrossRef]
31. Amitrano, C.; Vitale, E.; De Micco, V.; Arena, C. Light fertilization affects growth and photosynthesis in mung bean (*Vigna radiata*) plants. *J. Environ. Account. Manag.* **2018**, *6*, 295–304. [CrossRef]
32. Li, Q.; Kubota, C. Effects of supplemental light quality on growth and phytochemicals of baby leaf lettuce. *Environ. Exp. Bot.* **2009**, *67*, 59–64. [CrossRef]
33. Hernández, R.; Kubota, C. Physiological responses of cucumber seedlings under different blue and red photon flux ratios using LEDs. *Environ. Exp. Bot.* **2016**, *121*, 66–74. [CrossRef]
34. Sams, C.E.; Kopsell, D.; Morrow, R.C. Light quality impacts on growth, flowering, mineral uptake and petal pigmentation of marigold. Proc. *VIII Int. Symp. Light Hortic.* **2016**, *8*, 139–145.
35. Gill, S.S.; Tuteja, N. Reactive oxygen species and antioxidant machinery in abiotic stress tolerance in crop plants. *Plant Physiol. Biochem.* **2010**, *48*, 909–930. [CrossRef]
36. Tuteja, N.; Gill, S.S.; Tuteja, R. Plant responses to abiotic stresses: Shedding light on salt, drought, cold and heavy metal stress. In *Omics and Plant Abiotic Stress Tolerance*; Tuteja, N., Singh, S.G., Tuteja, R., Eds.; Bentham Science Publisher Ltd.: Sharjah, United Arab Emirates, 2011; Volume 3, pp. 39–64.
37. Bian, Z.H.; Yang, Q.C.; Liu, W.K. Effects of light quality on the accumulation of phytochemicals in vegetables produced in controlled environments: A review. *Sci. Food Agric.* **2015**, *95*, 869–877. [CrossRef]
38. Ohashi-Kaneko, K.; Matsuda, R.; Goto, E.; Fujiwara, K.; Kurata, K. Growth of rice plants under red light with or without supplemental blue light. *Soil Sci. Plant Nutr.* **2006**, *52*, 444–452. [CrossRef]
39. Hasan, M.M.; Bashir, T.; Gosh, R.; Lee, S.K.; Bae, H. An Overview of LEDs' Effects on the Production of Bioactive Compounds and Crop Quality. *Molecules* **2017**, *12*, 1420. [CrossRef]
40. Massa, G.D.; Kim, H.H.; Wheeler, R.M.; Mitchell, C.A.; Cary, A. Plant productivity in response to LED lighting. *Hortic. Sci.* **2008**, *43*, 1951–1956. [CrossRef]
41. Avercheva, O.; Berkovich, Y.A.; Smolyanina, S.; Bassarskaya, E.; Pogosyan, S.; Ptushenko, V.; Erokhin, A.; Zhigalova, T. Biochemical, photosynthetic and productive parameters of *Chinese cabbage* grown under blue–red LED assembly designed for space agriculture. *Adv. Space Res.* **2014**, *53*, 1574–1581. [CrossRef]
42. Mitchell, C.A.; Dzakovich, M.P.; Gomez, C.; Lopez, R.; Burr, J.F.; Hernández, R.; Kubota, C.; Currey, C.J.; Meng, Q.; Runkle, E.S.; et al. Light-emitting diodes in horticulture. *Hortic. Rev.* **2015**, *43*, 1–87.
43. Gómez, C.; Izzo, L.G. Increasing efficiency of crop production with LEDs. *AIMS Agric. Food* **2018**, *3*, 135–153. [CrossRef]
44. De Micco, V.; Amitrano, C.; Vitaglione, P.; Ferracane, C.; Pugliese, M.; Arena, C. Effect of light quality and ionising radiation on morphological and nutraceutical traits of sprouts for astronauts' diet. *Acta Astronaut.* **2021**, *185*, 188–197. [CrossRef]
45. Vitale, E.; Vitale, L.; Costanzo, G.; Velikova, V.; Tsonev, T.; Simoniello, P.; De Micco, V.; Arena, C. Light Spectral Composition Influences Structural and Eco-Physiological Traits of *Solanum lycopersicum* L. cv. 'Microtom' in Response to High-LET Ionizing Radiation. *Plants* **2021**, *10*, 1752. [CrossRef] [PubMed]
46. Ivanović, L.; Milašević, I.; Topalović, A.; Durović, D.; Mugoša, B.; Knežević, M.; Vrvić, M. Nutritional and phytochemical content of Swiss chard from Montenegro, under different fertilization and irrigation treatments. *Br. Food J.* **2018**, *121*, 411–425. [CrossRef]
47. Sjahril, R.; Riadi, M.; Raffiundin, S.T.; Toriyama, K.; Abe, T.; Trisnawaty, A.R. Effect of heavy ion beam irradiation on germination of local Toraja rice seed (M1-M2) mutant generation. IOP Conference Series. *Earth Environ. Sci.* **2018**, *157*, 012046. [CrossRef]
48. Komai, F.; Shikazono, N.; Tanaka, A. Sexual modification of female spinach seeds (*Spinacia oleracea* L.) by irradiation with ion particles. *Plant Cell Rep.* **2003**, *21*, 713–717. [CrossRef]
49. Hammond, E.C.; Bridgers, K.; Berry, F.D. Germination, growth rates, and electron microscope analysis of tomato seeds flown on the LDEF. *Radiat. Meas.* **1996**, *26*, 851–861. [CrossRef]

50. Barrero, J.M.; Downie, A.B.; Xu, Q.; Gubler, F. A Role for Barley Cryptochrome1 in Light Regulation of Grain Dormancy and Germination. *Plant Cell* **2014**, *26*, 1094–1104. [CrossRef]
51. Shinomura, T.; Nagatani, A.; Hanzawa, H.; Kubota, M.; Watanabe, M.; Furuya, M. Action spectra for phytochrome A-and B-specific photoinduction of seed germination in *Arabidopsis thaliana*. *Proc. Natl. Acad. Sci. USA* **1996**, *93*, 8129–8133. [CrossRef]
52. Thiede, M.E.; Link, S.O.; Fellows, R.J.; Beedlow, P.A. Effect of gamma radiation on stem diameter growth, carbon gain and biomass partitioning in Helianthus annus. *Environ. Exp. Bot.* **1995**, *35*, 33–41. [CrossRef]
53. Honda, I.; Kikuchi, K.; Matsuo, S.; Fukuda, M.; Saito, H.; Ryuto, H.; Fukunishi, N.; Abe, T. Heavy-ion-induced mutants in sweet pepper isolated by M1 plant selection. *Euphytica* **2006**, *152*, 61–66. [CrossRef]
54. De Micco, V.; Paradiso, R.; Aronne, G.; De Pascale, S.; Quarto, M.; Arena, C. Leaf Anatomy and Photochemical Behaviour of *Solanum lycopersicum* L. Plants from Seeds Irradiated with Low-LET Ionising Radiation. *Sci. World J.* **2014**, *2014*, 428141. [CrossRef] [PubMed]
55. Jo, Y.D.; Kim, S.H.; Hwang, J.E.; Kim, Y.S.; Kang, H.S.; Kim, S.W.; Kwon, S.J.; Ryu, J.; Kim, J.B.; Kang, S.Y. Construction of mutation populations by Gamma-ray and Carbon beam irradiated in Chili pepper (*Capsicum annuum* L.). *Hortic. Environ. Biotchnol.* **2016**, *57*, 606–614. [CrossRef]
56. Kim, H.H.; Wheeler, R.M.; Sager, J.C.; Gains, G.D.; Naikane, J.H. Evaluation of lettuce growth using supplemental green light with red and blue light-emitting diodes in a controlled environment—A review of research at Kennedy Space Center. V International Symposium on Artificial Lighting in Horticulture. *Acta Hortic.* **2006**, *711*, 111–120. [CrossRef]
57. Ursino, D.J.; Schefski, H.; McCabe, J. Radiation-induced changes in rate of photosynthetic CO_2 uptake in soybean plants. *Environ. Exp. Bot.* **1977**, *17*, 27–34. [CrossRef]
58. Jia, C.F.; Li, A.L. Effect of gamma radiation on mutant induction of *Fagopyrum dibotrys* Hara. *Photosynthetica* **2008**, *46*, 363–369. [CrossRef]
59. Moghaddam, S.S.; Jaafar, H.; Ibrahim, R.; Rahmat, A.; Maheran, A.A.; Philip, E. Effects of Acute Gamma Irradiation on Physiological Traits and Flavonoid accumulation of *Centella asiatica*. *Molecules* **2011**, *16*, 4994–5007. [CrossRef]
60. Fan, J.; Shi, M.; Huang, J.Z.; Xu, J.; Wang, Z.D.; Guo, D.P. Regulation of photosynthetic performance and antioxidant capacity by 60Co γ-irradiation in *Zizania latifolia* plants. *J. Environ. Radioact.* **2014**, *129*, 33–42. [CrossRef]
61. Jones, H. What is water use efficiency. In *Water Use Efficiency in Plant Biology?* Bacon, M., Ed.; Blackwell Publishing: Hoboken, NJ, USA; CRC Press: Boca Raton, FL, USA, 2004.
62. Sun, J.; Nishio, J.N.; Vofelmann, T.C. Green light drives CO_2 fixation deep within leaves. *Plant Cell Physiol.* **1998**, *39*, 1020–1026. [CrossRef]
63. Brodersen, C.R.; Vogelmann, T.C. Do changes in light direction affect absorption profiles in leaves? *Funct. Plant Biol.* **2010**, *37*, 403–412. [CrossRef]
64. Terashima, I.; Hanba, Y.T.; Tholen, D.; Niinemets, U. Leaf Functional Anatomy in Relation to Photosynthesis. *Plant Physiol.* **2011**, *155*, 108–116. [CrossRef] [PubMed]
65. Smith, H.L.; McAusland, L.; Murchie, E.H. Don't ignore the green light: Exploring diverse roles in plant processes. *J. Exp. Bot.* **2017**, *68*, 2099–2110. [CrossRef] [PubMed]
66. Zhang, L.; Van-Labeke, M.C. Long-term effects of Red-and Blue-Light Emitting Diodes on leaf anatomy and photosynthetic efficiency of three ornamental pot plants. *Front. Plant Sci.* **2017**, *8*, 917. [CrossRef] [PubMed]
67. Akoyunoglou, G.; Anni, H. Blue light effect on chloroplast development in higher plants. In *Blue Light Effects in Biological Systems*; Senger, H., Ed.; Springer: Berlin, Germany, 1984; pp. 397–406.
68. Sæbø, A.; Krekling, T.; Applegren, M. Light quality affects photosynthesis and leaf anatomy of birch plantlets in vitro. *Plant Cell Tissue Org. Cult.* **1995**, *41*, 177–185. [CrossRef]
69. Brown, C.S.; Schuerger, A.C.; Sager, J.C. Growth and photomorphogenesis of pepper plants under red light-emitting diodes with supplemental blue or far-red lighting. *J. Am. Soc. Hortic. Sci.* **1995**, *120*, 808–813. [CrossRef] [PubMed]
70. Yang, L.Y.; Wang, L.T.; Ma, J.H.; Ma, E.D.; Li, J.Y.; Gong, M. Effect of light quality on growth and development, photosynthetic characteristics and content of carbohydrates in tobacco (*Nicotiana tabacum* L.) plants. *Photosynthetica* **2017**, *55*, 467–477. [CrossRef]
71. Guidi, L.; Lo Piccolo, E.; Landi, M. Chlorophyll Fluorescence, Photoinhibition and Abiotic Stress: Does it Make Any Difference the Fact to Be a C3 or C4 Species? *Front. Plant Sci.* **2019**, *10*, 174. [CrossRef] [PubMed]
72. Demmig-Adams, B.; Garab, G.; Adams, W.W.; Govindjee, U. *Non-Photochemical Quenching and Energy Dissipation in Plants, Algae and Cyanobacteria*; Springer: Dordrecht, The Netherlands, 2014.
73. Hamdani, S.; Khan, N.; Perveen, S.; Qu, M.; Jiang, J.; Govindjee, U.; Zhu, X.G. Changes in the photosynthesis properties and photoprotection capacity in rise (*Oryza sativa*) grown under red, blue, or white light. *Photosynth. Res.* **2019**, *139*, 107–121. [CrossRef]
74. Sood, S.; Gupta, V.; Tripathy, B.C. Photoregulation of the greening process of wheat seedlings grown in red light. *Plant Mol. Biol.* **2005**, *59*, 269–287. [CrossRef]
75. Fan, X. Antioxidant capacity of fresh-cut vegetables exposed to ionizing radiation. *J. Sci. Food Agric.* **2005**, *85*, 995–1000. [CrossRef]
76. Liu, Y.; Wang, T.; Fang, S.; Zhou, M.; Qin, J. Responses of Morphology, Gas Exchange, Photochemical Activity of Photosystem II, and Antioxidant Balance in *Cyclocaryapaliurus* to Light Spectra. *Front. Plant Sci.* **2018**, *9*, 1704. [CrossRef]
77. Stajner, D.; Milosevic, M.; Popovic, B.M. Irradiation effects on phenolic content, lipid and protein oxidation and scavenger ability of soybean seeds. *Int. J. Mol. Sci.* **2007**, *8*, 618–627. [CrossRef]

78. Kiong, A.L.P.; Lai, A.G.; Hussein, S.; Harun, A.R. Physiological Responses of *Orthosiphon stamineus* Plantles to Gamma Irradiation. *Am. Eurasian J. Sustain. Agric.* **2008**, *2*, 135–149.
79. El-Beltagi, H.S.; Ahmed, O.K.; El-Desouky, W. Effect of low doses γ-irradiation on oxidative stress and secondary metabolites production of rosemary (*Rosmarinus officinalis* L.) callus culture. *Radiat. Phys. Chem.* **2011**, *80*, 968–976. [CrossRef]
80. Hammeed, A.; Shah, T.M.; Atta, B.M.; Haq, M.A.; Sayed, H. Gamma Irradiation effects on seed germination and growth, protein content, peroxidase and protease activity, lipid peroxidation in Desi and Kabuli Chickpea. *Pak. J. Bot.* **2008**, *40*, 1033–1041.
81. Chen, Q.F.; Ya, H.Y.; Feng, Y.R.; Jiao, Z. Expression of the Key Genes Involved in ABA Biosynthesis in Rice Implanted by Ion Beam. *Appl. Biochem. Biotechnol.* **2014**, *173*, 239–247. [CrossRef]
82. Zhang, L.; Liu, S.; Zhang, Z.; Yang, R.; Yang, X. Dynamic effects of different light qualities on pea sprouts quality. *North. Hortic.* **2010**, *8*, 4–7.
83. Li, H.; Tang, C.; Xu, Z.; Liu, X.; Han, X. Effects of Different Light Sources on the Growth of Non-heading Chinese Cabbage (*Brassica campestris* L.). *J. Agric. Sci.* **2012**, *4*, 262–273. [CrossRef]
84. Lattanzio, V.; Kroon, P.A.; Quideau, S. Plant phenolics—Secondary metabolites with diverse functions. In *Recent Advances in Polyphenol Research*; Daay, F., Lattanzio, V., Eds.; Wiley-Blackwell: Oxford, UK, 2008; pp. 1–35.
85. De Micco, V.; Arena, C.; Aronne, G. Anatomical alterations of *Phaseolus vulgaris* L. mature leaves irradiated with X-rays. *Plant Biol.* **2014**, *16*, 187–193. [CrossRef]
86. He, J.; Dong, L.; Lixia, Y.; Wenjian, L. Pigment analysis of color-leaf mutant in Wanderig Jew (*Tradescantiafluminensis*) irradiated by carbon ions. *Nucl. Sci. Techol.* **2011**, *22*, 77–83.
87. Arena, C.; De Micco, V.; Aronne, G.; Pugliese, M.; Virzo, A.; De Maio, A. Response of *Phaseolus vulgaris* L. plants to low-LET ionisingradiation: Growth and oxidative stress. *Acta Astronaut.* **2013**, *91*, 107–114. [CrossRef]
88. Lekkham, P.; Srilaong, V.; Pongprasert, N.; Kondo, S. Anthocyanin concentration and antioxidant activity in light-emitting diode (LED)-treated apples in a greenhouse environmental control system. *Fruits* **2016**, *71*, 269–274. [CrossRef]
89. Agarwal, A.; Gupta, S.D.; Barman, M.; Mitra, A. Photosynthetic apparatus plays a central role in photosensitive physiological acclimations affecting spinach (*Spinacia oleracea* L.) growth in response to blue and red photon flux ratios. *Environ. Exp. Bot.* **2018**, *156*, 170–182. [CrossRef]
90. Lobiuc, A.; Vasilache, V.; Pintilie, O.; Stoleru, T.; Burducea, M.; Oroian, M.; Zamfirache, M.M. Blue and Red LED Illumination Improves Growth and Bioactive Compounds Content in Acyanic and Cyanic *OciumBasilicum* L. Microgreens. *Molecules* **2017**, *22*, 2111. [CrossRef] [PubMed]
91. Livadariu, O.; Raiciu, D.; Maximilian, C.; Capitanu, E. Studies regarding treatments of LED-s emitted light on sprouting hemp (*Cannabis sativa* L.). *Rom. Biotechnol. Lett.* **2018**, *24*, 485–490. [CrossRef]
92. Alothman, M.; Bhat, R.; Karim, A.A. Effects of radiation processing on phytochemicals and antioxidants in plant procedure. *Trends Food Sci. Technol.* **2009**, *20*, 201–212. [CrossRef]
93. Sallam, E.M.; Anwar, M.M. Antioxidant Activity of some Extracts from Gamma irradiated Purslane (*Portulaca oleracea*) Plant. *Int. J. Agric. Biol.* **2017**, *19*, 48–52. [CrossRef]
94. Ohashi-Kaneko, K.; Takase, M.; Kon, N.; Fukiwara, K.; Kurata, K. Effect of light quality in leaf lettuce, spinach and komatsuna. *Environ. Control Biol.* **2007**, *45*, 189–198. [CrossRef]
95. Samuolienė, G.; Urbanovičiūtė, A.; Brazaitytė, A.; Šabajevienė, G.; Sakalauskaitė, J.; Duchovskis, P. The impact of LED illumination on antioxidant properties of sprouted seeds. *Cent. Eur. J. Biol.* **2011**, *6*, 68–74. [CrossRef]
96. Ninfali, P.; Angelino, D. Nutritional and functional potential of *Beta vulgariscicla* and rubra. *Fitoterapia* **2013**, *89*, 188–199. [CrossRef]
97. Kazama, Y.; Hirano, T.; Saito, H.; Liu, Y.; Ohbu, S.; Hayashi, Y.; Abe, T. Characterization of highly efficient heavy-ion mutagenesis in Arabidopsis thaliana. *Plant Biol.* **2011**, *11*, 161. [CrossRef] [PubMed]
98. von Caemmerer, S.; Farquhar, G.D. Some relationships between the biochemistry of photosynthesis and the gas exchange of leaves. *Planta* **1981**, *153*, 376–387. [CrossRef] [PubMed]
99. Genty, B.; Briantais, J.M.; Baker, N.R. The relationship between the quantum yield of photosynthetic electron transport and quenching of chlorophyll fluorescence. *Biochim. Biophys. Acta* **1989**, *990*, 87–92. [CrossRef]
100. Bilger, W.; Björkman, O. Temperature dependence of violaxanthin deepoxidation and nonphotochemical fluorescence quenching in intact leaves of *Gossypium kirsutum* L. and *Malva parviflora* L. *Planta* **1991**, *184*, 226–234. [CrossRef]
101. Lichtentaler, H.K. Chlorophylls and Carotenoids: Pigments of photosynthetic biomembranes. *Meth. Enzymol.* **1987**, *148*, 350–382.
102. Mancinelli, A.L.; Yang, C.P.H.; Lindquist, P.; Anderson, O.R.; Rabino, I. Photocontrol of Anthocyanin Synthesis. *Plant Physiol.* **1975**, *55*, 251–257. [CrossRef]
103. George, B.; Kaur, C.; Khurdiya, D.S.; Kapoor, H.C. Antioxidants in tomato (*Lycopersicumesculentum*) as a function of genotype. *Food Chem.* **2004**, *84*, 45–51. [CrossRef]
104. Hedge, J.E.; Hofreiter, B.T. Carbohydrate chemistry. In *Methods in Carbohydrate Chemistry*; Whistler, R.L., Be Miller, J.N., Eds.; Academic Press: New York, NY, USA, 1962; Volume 17.
105. Wang, W.; Vignani, R.; Scali, M.; Cresti, M. A universal and rapid protocol for protein extraction from recalcitrant plant tissue for proteomic analysis. *J. Electrophor.* **2006**, *27*, 2782–2786. [CrossRef]
106. Bradford, M.M. A rapid and sensitive method for the quantitation of microgram quantities of protein utilizing the principle of protein-dye binding. *Anal. Biochem.* **1976**, *72*, 248–254. [CrossRef]

plants

MDPI

Article

The Effect of LED Light Spectra on the Growth, Yield and Nutritional Value of Red and Green Lettuce (*Lactuca sativa*)

Abdullah A. Alrajhi [1,*], Abdulaziz S. Alsahli [1,2], Ibrahim M. Alhelal [2], Hail Z. Rihan [3,4,*], Michael P. Fuller [3], Abdullah A. Alsadon [5] and Abdullah A. Ibrahim [5]

1 National Center for Agriculture Technology, King Abdulaziz City for Science and Technology, P.O. Box 6086, Riyadh 11442, Saudi Arabia
2 Department of Agricultural engineering, College of Food and Agriculture Sciences, King Saud University, P.O. Box 2460, Riyadh 11451, Saudi Arabia
3 School of Biological and Marine Sciences, Faculty of Science and Engineering, University of Plymouth, Plymouth PL4 8AA, UK
4 Phytome Life Sciences, Launceston PL15 7AB, UK
5 Department of Plant Production, College of Food and Agriculture Sciences, King Saud University, P.O. Box 2460, Riyadh 11451, Saudi Arabia
* Correspondence: aalrajhi@kacst.edu.sa (A.A.A.); hail.rihan@plymouth.ac.uk (H.Z.R.)

Abstract: Controlled Environment Agriculture (CEA) is a method of increasing crop productivity per unit area of cultivated land by extending crop production into the vertical dimension and enabling year-round production. Light emitting diodes (LED) are frequently used as the source of light energy in CEA systems and light is commonly the limiting factor for production under CEA conditions. In the current study, the impact of different spectra was compared with the use of white LED light. The various spectra were white; white supplemented with ultraviolet b for a week before harvest; three combinations of red/blue lights (red 660 nm with blue 450 nm at 1:1 ratio; red 660 nm with blue 435 nm 1:1 ratio; red 660 nm with blue at mix of 450 nm and 435 nm 1:1 ratio); and red/blue supplemented with green and far red (B/R/G/FR, ratio: 1:1:0.07:0.64). The growth, yield, physiological and chemical profiles of two varieties of lettuce, Carmoli (red) and Locarno (green), responded differently to the various light treatments. However, white (control) appeared to perform the best overall. The B/R/G/FR promoted the growth and yield parameters in both varieties of lettuce but also increased the level of stem elongation (bolting), which impacted the quality of grown plants. There was no clear relationship between the various physiological parameters measured and final marketable yield in either variety. Various chemical traits, including vitamin C content, total phenol content, soluble sugar and total soluble solid contents responded differently to the light treatments, where each targeted chemical was promoted by a specific light spectrum. This highlights the importance of designing the light spectra in accordance with the intended outcomes. The current study has value in the field of commercial vertical farming of lettuce under CEA conditions.

Keywords: controlled environment agriculture (CEA); light emitting diodes (LED); light spectra; lettuce; yield; photosynthetic rate; vitamin C

Citation: Alrajhi, A.A.; Alsahli, A.S.; Alhelal, I.M.; Rihan, H.Z.; Fuller, M.P.; Alsadon, A.A.; Ibrahim, A.A. The Effect of LED Light Spectra on the Growth, Yield and Nutritional Value of Red and Green Lettuce (*Lactuca sativa*). *Plants* **2023**, *12*, 463. https://doi.org/10.3390/plants12030463

Academic Editor: Fermin Morales

Received: 18 October 2022
Revised: 16 January 2023
Accepted: 16 January 2023
Published: 19 January 2023

1. Introduction

Environmental conditions present a major limiting factor for food production in most regions of the world and this is particularly acute in desert countries such as Saudi Arabia and most of the Middle East. High temperatures and water availability are major limiting factors in these regions, and intense irradiation often exceeds crop absorption capacity. As a consequence, "Controlled Environment Agricultural" (CEA) systems, or what is commonly called "Plant Factory" (PF) systems, could provide promising solutions for food security in these regions. Protected agriculture has developed from simple polytunnel structures to high-tech glasshouses and is now developing to completely CEA that optimizes the productivity and quality of plants. Major technical developments influenced by advances in precision technology, lighting, automation and data processing are making CEA systems a cost-efficient reality [1]. CEA systems offer greater predictability and increases in crop yield per unit area together with improvements in crop quality and improved shelf-life.

A PF is a typical vertical farming model based on a high-rise multi-level factory design where recycled water is supplemented with nutrients and used to feed plants through an advanced hydroponic system [2]. Light emitting diodes (LED) lighting systems are used as the sole source of light (energy) in the PF setting and the energy consumption of the PF can be offset using sustainable power generating systems such as photo-voltaics or wind turbines. Such PFs are being created as both research and production facilities.

Light has a great impact on the growth, yield, development, morphology and chemical composition of plants [3]. Light intensity (radiant density), light spectrum and photoperiod all have significant impacts on plant growth and physiology [4]. LED lighting systems have unique features that enable them to be optimized for PF settings, including high light-use efficiency, long life span and high efficacy, leading to reduced associated heating [5] compared to other artificial lighting sources, such as fluorescent tubes or sodium vapor lamps. Furthermore, spectral specificity can be introduced through the design of the LED array by employing a mixture of LEDs with different wavelength emittance, controlled through appropriate control systems [6–8]. Such systems have a high degree of commercial applicability.

Plant species react differently to various wavelengths due to differences in their photoreceptors [9,10]. A significant amount of research has investigated the impact of LEDs on the growth, shape, yield and edible quality parameters of several plant species [6,7,11–14]. The impact of LEDs on the chemical composition, such as vitamin C content, chlorophyll content, soluble sugar, protein level and antioxidant activity has also been the subject of an increasing amount of recent research [15–17].

Both red (R) and blue (B) wavelengths are reported to have significant impacts on plant growth and development. These wavelengths are mainly absorbed by photosynthetic pigments, leading to carbon assimilation. However, these wavelengths can also have major effects on plant architecture and development [4]. Red induces transformations in phytochrome and is important for phytochemical synthesis, such as phenolics and oxalate. Red wavelengths are also essential for the development of the photosynthetic apparatus [18,19]. Blue is crucial for the development of chloroplasts and for photomorphogenesis. Blue wavelengths are important in influencing stomatal opening and for the upregulation of chlorophyll and anthocyanin [20,21]. Green wavelengths, although not readily absorbed by the red/blue photosynthetic pigments, have an impact on the growth and development of several plants and it has been reported that the addition of green light to the red/blue LED arrays has a positive and significant impact on the leaf growth of lettuce and on photosynthetic activities [22,23]. Legendre and Van Lersel [24] reported the important impact of far-red on plant architecture and its positive influence on plant photosynthesis. Wavelengths shorter than blue fall into the UV range and have been reported to have a positive effect on the accumulation of secondary metabolites in plants. These secondary metabolites often accumulate and perform many functions, including acting as sunscreens in intense radiation environments and chemical protectants against insect herbivory. Ferreyra et al., for example [25], reported that flavonoids were accumulated under the impact of UV-B radiation.

Lettuce is widely cultivated in plant factories under LEDs [26] because of its capacity to adapt to controlled environments. It is often used as a "test crop" in new installations. It has a short growth cycle and a defined rosette shoot shape [27,28]. Several researchers have investigated the impact of light spectra on the growth and development of lettuce [29] and have reported that combined red and blue LEDs exhibit the highest chlorophyll content and photosynthesis rate. However, when monochromatic blue or red LEDs were applied alone, lettuce plants showed growth abnormality and a decrease in the rate of photosynthesis [30]. This finding is in agreement with what was reported by [31], indicating that the use of red LED alone resulted in reduction in biomass, chlorophyll content, carotenoid content and antioxidant levels in lettuce. Others have also reported that red light alone led to abnormality in lettuce, and a combination of 90% red and 10% blue had a positive impact compared to red light only [32]. Furthermore, it has been reported that an increased fraction of blue light at 435 nm in combination with red light at 663 nm at a high irradiance improved the physiological indicators and enhanced the yield of lettuce [33]. It was also reported that the photosynthetic rate and stomatal density depends on the red light/blue light ratio [29]. In this regard, it was reported that increasing R/B ratio from 0.5 to 3 enhanced the level of chlorophyll and flavonoid content, nutrient uptake and water use efficiency of lettuce [5].

The study reported here presents the results of the first experiment to be carried out in a customized PF research installation in Saudi Arabia. The PF was designed by researchers from the University of Plymouth UK and assembled in China and was constructed in a shipping container. The experiment aimed to investigate the impact of a wide range of LED light spectra, including red, blue, green, far red, UV and white on the growth, yield and quality traits of two varieties of lettuce (green and red). Moreover, it investigated whether the relative ratio of blue and red LEDs would impact the growth, physiology and chemical profile of lettuce compared to a full white spectrum. Finally, the impact of supplementing the blue/red growth lights with UV, far-red and green LEDs was also investigated. According to our data, this is the first study to investigate the impact of two types of blues (450 nm and 435 nm) on the growth, physiology and chemistry of two types of lettuces. Therefore, the current paper aimed to provide a deeper understanding of the impact of light spectra on lettuce growth and development.

2. Materials and Methods

Seeds of two varieties of lettuce, Carmoli (red lettuce) and Locarno (green lettuce) (Rijk Zwaan, Burgemeester Crezéelaan, De Lier, The Netherlands) were sown and germinated for 14 days in the greenhouse at Muzahimiyah Research Station, King Abdulaziz City for Science and Technology (KACST). Seeds were sown in Rockwool cubes (35 mm) (one seed per cube) and when seedlings produced their first pair of true leaves, they were transferred to the KACST Plant Factory research facility. The KACST Plant Factory facility is a converted insulated modular shipping container where external light has been excluded. Its multi-tier hydroponic growing system consists of gullies for a nutrient film technique (NFT) hydroponic system and is installed with interchangeable LED light units. VitaLink Hydro MAX (HydroGarden, Coventry, UK) nutrients (highly-concentrated A and B formulations) was used as the nutrient solution in the system. The KACST Plant Factory system is divided into several multi-shelf units, each consisting of four tiers 50 cm apart and plants spaced 15 cm apart. The facility was fitted with air conditioning and dehumidification systems and the temperature and humidity were monitored using an Autogrow control system (Autogrow Systems Ltd., Auckland, New Zealand) set at 22 ± 2 °C and 75 ± 5%, respectively. The dark/light period was set at 8/16 h. The facility is fitted with an entry "air shower" and operatives wore clean sterile overalls, shoes and hats to maintain a clean growing environment. No diseases or pests were observed throughout the experiments.

Light intensity (radiant density) was set at 250 ± 10 μmol m^{-2} s^{-1} and each LED wavelength was adjustable so that each treatment maintained this light intensity. Six Lumnigrow lighting arrays were used (Table 1) (LuminiGrow, Shenzhen, China) as follows:

1. White (50% cool white + 50% warm light) (W);
2. Red spectral bands with the maximum at660 nm and blue spectral bands with the maximum at 450 nm with ratio (1:1) (R/B450(1:1));
3. Red spectral bands with the maximum at 660 nm and blue spectral bands with the maximum at 435 with ratio (1:1), (R/B435(1:1));
4. Red spectral bands with the maximum at 660 nm and blue spectral bands with the maximum at 450, blue: red (1:1) supplemented by far red and green with ratio (1:1:0.07:0.64) (B/R/G/FR (1:1:0.07:0.64). Green is a wide range wavelength (500–600 nm). Rare red spectral bands with the maximum at 725 nm;

5. White (50% cool white + 50% warm light) + UV-B (0.3 μW cm^{-1}) (for an hour/24 h for a week before harvest) (W/UV-B). UV-B intensity was set at 0.3 microwatt.cm^{-2} measured using UV Light Meter UV340B (UV Light Meter, UV340B, Shenzhen Ever Good Electronic Co., Ltd., Shenzhen City, China). The UV-B was set to switch on during the lighting period of the other wavelengths;
6. Red spectral bands with the maximum at 660 nm and blue at combination of blue spectral bands with the maximum at 435 nm + 450 nm with ratio (1:1), blue: red (1:1). (R/B450-435 (1:1) (Figure 1).

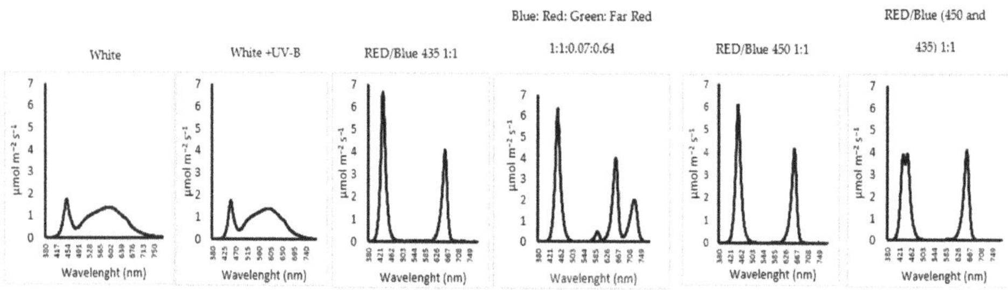

Figure 1. Spectra of the lighting treatments used, as measured by a UPRtek spectrophotometer (Zhunan Township, Miaoli County 35059, Taiwan).

Table 1. The relative ratio of light spectrum in the applied treatments (R refers to red, B refers to blue, W refers to white, G refers to green, FR refers to far-red and UV to ultraviolet.

Treatment	Red (660 nm)	Blue (450 nm)	Blue (435 nm)	Green (520 nm)	Far Red	UV-B
R/B450(1:1)	1	1	-	-	-	-
W	0.94	0.53		1	0.13	-
R/B435(1:1)	1	-	1	-	-	-
B/R/G/FR	1	1	-	0.07	0.64	-
W/UV-B	0.94	0.53		1	0.13	+
R/B450-435	1		1		-	-

2.1. Morphological Measurements

Morphological measurements included plant height (cm); leaf area (LA) (mm^2) measured using an AM350 Portable Leaf Area Meter (ADC BioScientific, Herts, UK); shoot fresh weight (FW); and dry weight (DW) (g) (after removing the root system) using a sensitive Fisher Scientific SG-402 laboratory balance (Hampton, NH, USA). For dry weight, plants were dried at 70 °C for 96 h [34].

For each stage of development, three replicates (in area) of four plants for each treatment were assessed.

2.2. Physiological Measurements

Physiological responses to lighting treatments were measured at three stages of development: 15 days, 30 days and 45 days (final harvest stage) after planting. Physiological measurements included light-saturated instantaneous maximum photosynthetic rate Amax (μmol CO_2 m^{-2} s^{-1}) and stomatal conductance (mol H_2O m^{-2} s^{-1}) measured using an LI-COR 6400 Highly Portable Ambient Photosynthesis System (LI-COR LI 6400XT, Lincoln, NE, United States).

For each stage of development, three replicates of four plants for each treatment were assessed.

2.3. Chemical Analysis

Ascorbic acid (vitamin C) was estimated using High Performance Liquid Chromatography (HPLC) (Alliance Waters w2695 Separations Module, Milford, MA, USA). The extraction solution was prepared using 3 g of metaphosphoric acid added to 100 mL of distilled water. A total of 3 g of the plant fresh leaf material was added to the prepared extraction solution, and this was mixed using Glas/Gol vortex (Glas/Gol tools, 711 Hulman St., Terre Haute, IN, USA) at a speed of 70 RPM. The mix was transferred to a water bath (Branaon 8000 West Florissant Avenue, St. Louis, MO, USA) held at 30 °C. The plant extract was then filtered using filter paper (0.45 µm) and the supernatant used for the HPLC analysis using the following column and settings: NewcrcmBH (3.2 × 50 mm, 3 µm), mobile phase: $MeCN/H_2O$—50/50%, buffer: H_3PO_4—0.5%, flow rate: 0.5 mL min^{-1}, UV detection: 275 nm.

Total phenol content was estimated using gas chromatography (GC/MSD Systems—5975C Series GC/MSD System, 5301 Stevens Creek Blvd., Santa Clara, CA, USA). Helium was used as the carrier gas at an average flow rate of 28 cm^3 min^{-1}. A total of 5 g of plant material was ground using a pestle and mortar, 20 mL of methanol was added and the mix was filtered through a 125 mm filter paper. The supernatant obtained was then centrifuged at 1000 rpm for 10 min (Hermle Labortechnik GmbH, Siemensstr. 25, Wehingen, Germany).

HPLC was used for the estimation of total soluble sugar content. The column and settings were shim-pact CLC-NH_2 (6.0 MM i.d. × 15 cm), mobile phase: Acetonitrile/water (7/3), flow rate: 1.0 mL min^{-1}.A total of 5 g of plant fresh material was ground using a pestle ad mortar. A total of 20 mL of distilled water and 20 mL of ethanol were added to the plant material. The mix was filtered through a 125 mm filter. The supernatant obtained was then centrifuged at 1000 rpm for 10 min (Hermle Labortechnik GmbH, Siemensstr. 25, Wehingen, Germany) and used for the soluble sugar analysis.

The total soluble solids of lettuce juice were estimated using a digital portable refractometer (101 model, ATAGO, Japan).

For Chlorophyll estimation, 1 g of plant leaves were ground in 20 mL of 80% acetone using a pestle and mortar. Samples were then centrifuged for 5 min at 1000 rpm min^{-1} in a ROTOFIX 32 (Tuttlingen, Germany). The supernatant (2 mL) was placed in a cuvette and the absorbance was measured at 663.6 (A663.6) and 646.6 (A646.6) using Jenway 7315 (Staffordshire, UK). The formulae are based on the absorbance maxima of each pigment and are dependent on the solvent used. The formulae for samples dissolved in acetone are as follows:

$$Ca = 12.25\ A_{663.6} - 2.55\ A_{646.6},$$

$$Cb = 20.31\ A_{646.6} - 4.91\ A_{663.6},$$

where

$A_{663.6}$ is the solution absorbance at 663.6 nm and $A_{646.6}$ is the absorption at 646.6;

Ca: chlorophyll A, Cb: chlorophyll B, Total C: total chlorophyll.

The values obtained were converted to estimate the chlorophyll content (mg) per gram of fresh weight, following a previously described procedure [30].

Three biological replicates, each consisting of 4 plants (experimental replicates), were applied for each chemical test.

2.4. Statistical Analysis

For each lettuce variety, three replicates for each treatment were applied. Each replicate consisted of 4 plants (12 plants in total across the three developmental stages).

Results are presented as means ± standard error (S.E.). All data were subjected to analysis of variance (ANOVA), using Minitab software (version 19) and comparisons of means were made using the least significant difference test (LSD) at 5% level of probability.

3. Results

3.1. Morphological Parameters

The two varieties of lettuce (green and red) used in this experiment responded differently to the various growing light spectra and are presented separately for ease of interpretation.

For green lettuce, all of the various light spectra were capable of producing acceptable marketable lettuce, but some spectra performed better than others. In terms of final yield, white light and white light + UV provided the best crops with most leaves, highest leaf area and greatest fresh weight (Figure 2). R/B450 supplemented by far-red and green (R/B/G/FR 1:1:0.07:0.64) treatments significantly increased plant height ($p \leq 0.001$) by increasing stalk length. This increase in stalk length indicates a premature flowering response and is an undesirable characteristic for marketable lettuce.

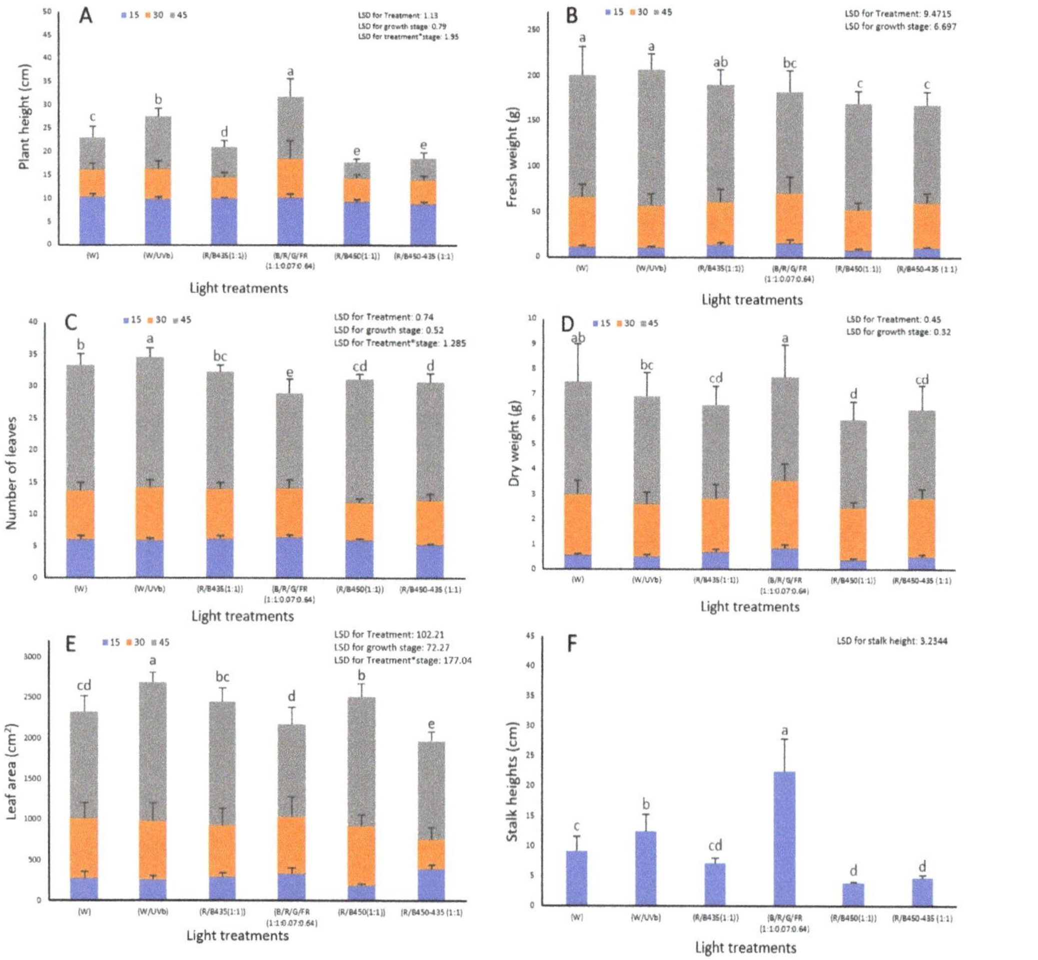

Figure 2. The impact of LED light spectrum on (**A**) plant height (cm); (**B**) plant fresh weight (g); (**C**) number of leaves; (**D**) plant fresh weight (g); (**E**) leaf area (cm^2) and (**F**) stalk height (cm) of green lettuce (Locarno variety) at three developmental stages, 15, 30 and 45 days from transplanting in the plant factory. Letters indicate significant differences ($p < 0.05$) between treatments within each experiment (* refers to the interaction between treatments).

These effects were evident even at early harvests (Figure 2).

For the red lettuce variety, the effects reported above for green lettuce were less accentuated. Whilst R/B450 supplemented by far-red and green (R/B/G/FR 1:1:0.07:0.64) treatments still led to increases in stalk length, the R/B/G/FR 1:1:0.07:0.64 treatment was the only treatment where this was at unacceptable levels for marketable lettuce. The white light treatment provided the most acceptable marketable lettuce.

Differences between the various light spectra were much less evident at the early developmental stages and growth rates significantly accelerated in the last 15 days of growth (between 30 and 45 days) (Figure 3).

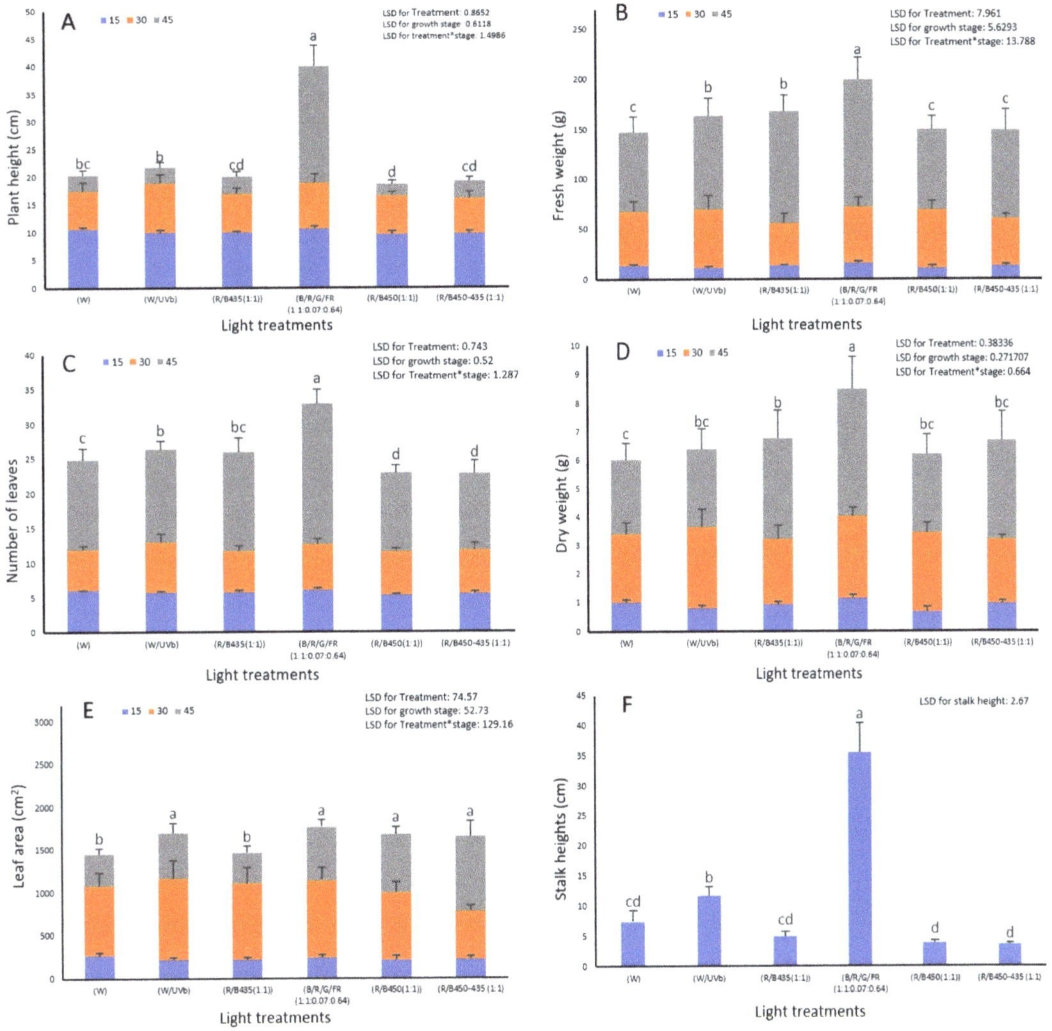

Figure 3. The impact of LED light spectrum on (**A**) plant height (cm); (**B**) plant fresh weight (g); (**C**) number of leaves; (**D**) plant fresh weight (g); (**E**) leaf area (cm^2) and (**F**) stalk height (cm) of Carmoli variety (red lettuce) at three developmental stages, 15, 30 and 45 days from transplanting in the plant factory. Letters indicate significant differences ($p < 0.05$) between treatments in each experiment.

3.2. Physiological Parameters

Light treatment had a significant impact on the photosynthetic rate of both green lettuce and red lettuce ($p \leq 0.001$) (Figure 4A). R/B 450 (1/1) and R/B 435 (1:1) resulted in plants with a higher photosynthetic rate in comparison with the use of the white spectrum. Overall, the photosynthetic rate in green lettuce was significantly higher than that of red lettuce ($p \leq 0.001$).

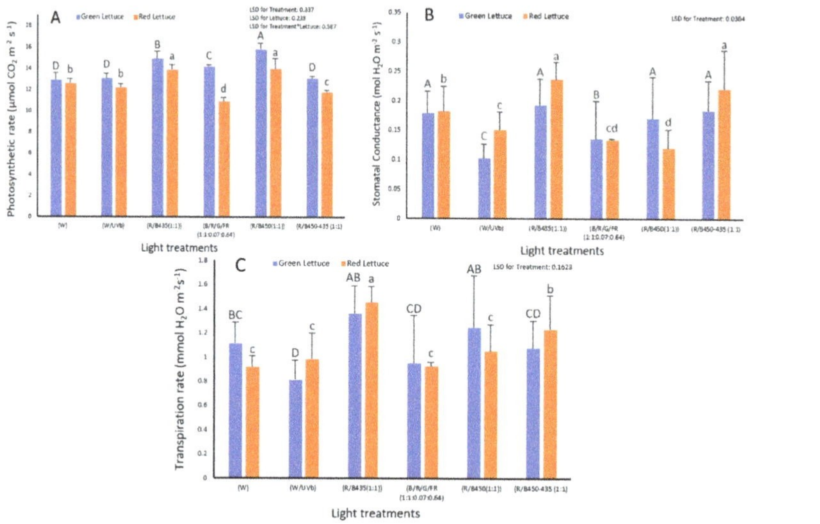

Figure 4. The impact of LED light spectrum on (**A**) photosynthetic rate (µmol CO_2 m^{-2} s^{-1}); (**B**) stomatal conductance (mol H_2O m^{-2} s^{-1}); (**C**) transpiration rate (mmol H_2O m^{-2} s^{-1}) of green lettuce (Locarno variety) Carmoli variety (red lettuce). Letters indicate significant differences ($p < 0.05$) between treatments in each experiment. (Capital letters are for green lettuce and small letters are for red lettuce.)

Light treatment had a significant impact on the stomatal conductance rate ($p \leq 0.001$). The highest level of stomatal conductance rate was observed when R/B 435 (1:1) light treatment was applied to both green and red lettuce (Figure 4B). There was impact of lettuce variety ($p = 0.305$) and no significant interaction between the lettuce varieties and light treatment ($p = 0.216$) (Figure 4B).

The highest level of transpiration rate was observed under R/B 435 (1:1) treatments ($p \leq 0.001$). However, there was no significant impact of the lettuce variety ($p = 0.998$) and no significant interaction between lettuce variety and light treatment on transpiration rate ($p = 0.446$) (Figure 4C).

3.3. Chemical and Quality Traits

A significant effect of light treatment ($p \leq 0.001$), lettuce variety ($p = 0.018$) and the interaction between light treatment and lettuce variety ($p \leq 0.001$) on the level of vitamin C was observed (Figure 5A). The highest level of vitamin C in green lettuce was observed under R/B/G/FR (1:1:0.07/0.64), whilst in red lettuce the highest level of vitamin C was observed under R/B 450 (1:1) and white light. Moreover, whilst the addition of UV-B treatment to white light seemed to have a significant negative impact on the content of vitamin C in red lettuce, the lowest level of this vitamin C was observed under R/B 450 (1:1) in green lettuce (Figure 5A).

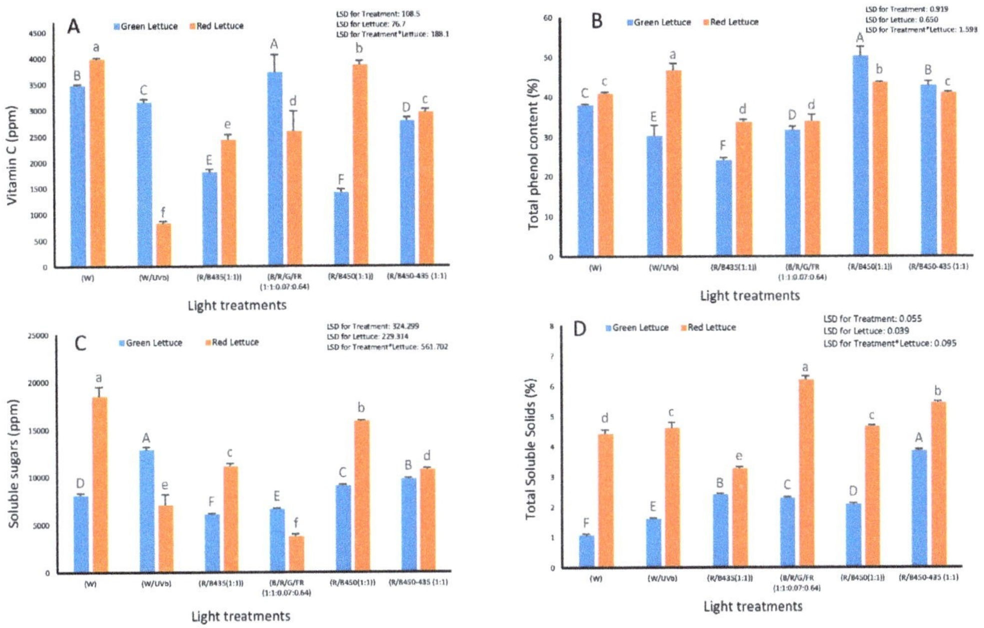

Figure 5. The impact of LED light spectrum on (**A**) vitamin C contents (PPM); (**B**) total phenol content (%); (**C**) soluble sugar content (PPM); (**D**) total soluble solids (%) of green lettuce (Locarno variety) Carmoli variety (red lettuce). Letters indicate significant differences ($p < 0.05$) between treatments within each experiment.

Total phenol content was also affected by light treatment, lettuce variety and the interaction between light treatment and lettuce variety ($p \leq 0.001$). The highest phenol content in green lettuce was observed using R/B 450 (1/1) whilst in red lettuce the content of phenol was the highest under white + UV-B treatment (Figure 5B).

There was a significant impact of light treatments, lettuce variety and the interaction between light treatment and lettuce variety ($p \leq 0.001$) on the level of total soluble sugar (Figure 5C). While white + UV light treatment induced the highest level of soluble sugar in green lettuce, growing red lettuce under the white light had the most positive impact (Figure 5C).

Total soluble solid was impacted by light treatment, lettuce variety and the interaction between light treatment and lettuce variety ($p \leq 0.001$). R:B450-435(1/1) introduced the highest level to total soluble solids in green lettuce. However, R:B:G:FR (1/1/0.07/0.64) light treatment induced the highest level of total soluble solids in red lettuce (Figure 5D).

Light spectrum had a significant impact on chlorophyll content in both green and red lettuce varieties ($p \leq 0.001$) and overall chlorophyll was lower in green lettuce than in red lettuce. R/B450 (1/1) induced the highest level of chlorophyll A and B in green lettuce (Figure 6A) and R/B/G/FR (1:1:0.07:0.64) provided the highest content in red lettuce (Figure 6B).

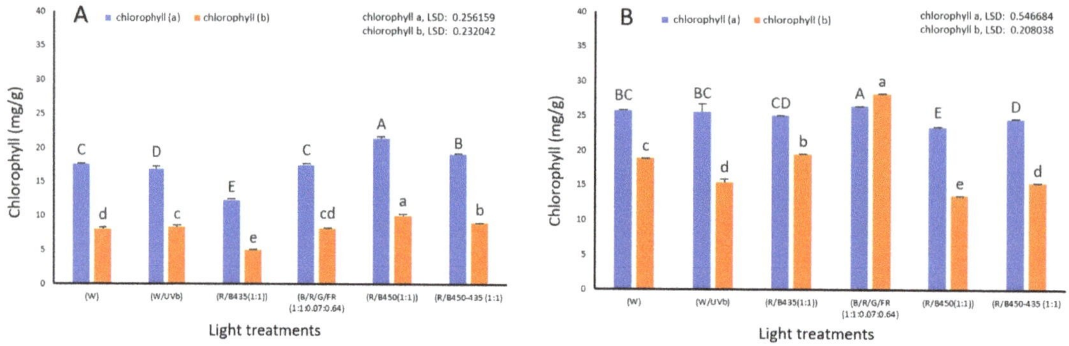

Figure 6. The impact of LED light spectrum on chlorophyll content of (**A**) blue lettuce and (**B**) red lettuce. Letters indicate significant differences ($p < 0.05$) between treatments within each experiment (capital letters for chlorophyll (**a**) and small letters for chlorophyll (**b**)).

4. Discussion

To obtain the maximum output from plants grown under CEA, particularly in commercial PF situations, lighting conditions need to be optimized. In particular, the light spectrum can now be manipulated using LED lighting arrays using customised research platforms similar to those installed in the KACST Plant Factory research facility [6,35]. It was shown in the current study that the light spectrum needs can even be variety-specific within a species, indicating that there can be differences in photoreceptors between commercial varieties, a finding supported by other published work [6–10,33].

White light in CEA or PFs is usually supplied using designated white LEDs, which tend to be phosphor-coated blue LEDs generating a full white spectrum including a green rich portion as shown in Figure 1. However, "white light" can also be supplied by employing a combination of monochromatic red, blue and blue LEDs in an array. Both approaches provide a white light and make the facility more user-friendly, but it is not clear which is the best to use for commercial production. Many recent commercial facilities have not opted for white light but instead have invested in red/blue lights to maximize the utilizable radiant energy produced per kW of electrical energy input (the efficacy). The results presented here for two varieties of lettuce clearly demonstrate that both white light and red/blue and red/blue/blue combinations are capable of producing marketable lettuce but significant differences did exist in growth and development rates. White light produced the best plants in terms of commercial yield at the end of the experiment, whereas R/B and R/B/G led to significant stalk lengthening, reducing the marketability. However, the results indicated that under R/B and R/B/G, marketable stage had actually been reached earlier, and although a smaller individual size may have occurred as a result, a marketable product had been achieved some days earlier compared to that under white light. This finding is in agreement with other research that reported the positive impact of white light on the growth and development of plants [36]. The current findings are also in accordance with other research, indicating that lettuce plants grown with spectra that included green light had better growth parameters, such as fresh and dry weights, compared to those grown with light focused in the red and blue region [37]. These findings with lettuce contradict the reports with other species such as basil, which indicate that focusing the lights in the red/blue spectrum promotes growth and yield [7,8], asserting that species can vary in their spectral needs. In our experiment, two R/B LED combinations were tested with the blue either at the 450 nm range corresponding to chlorophyll B absorption or 435 nm corresponding to chlorophyll A absorption as our previous research with basil had indicated a distinct preference for 435 nm over the more conventionally used 450 nm (9). In this case, the wavelength of blue at 435 nm did not promote the growth and yield, as well as the use of blue at 450 nm, which contradicts the findings obtained with basil (9).

These species differences appear to be genetically based, which emphasizes that different species respond differently to light spectra, probably due to differences in terms of pigment composition, photoreceptors, plant morphological structure and chemical composition.

Two supplementary non-photosynthetically active LED spectra were also investigated in the current experiment, UV-B and far-red. Whilst the UV-B had no effect on growth and production, the far red did. The far-red supplementation is becoming increasing in interest since Zou (2019) [38], who reported that the total biomass of lettuce was increased by 39% and 25% when the light spectrum was supplemented with FR-Day (through the day) and FR-EOD (end of the day) treatments, respectively. Far red is known to be involved in Phytochrome regulation which in turn regulates morphological and physiological properties and plays a critical role in mediating plant growth and development [39]. In the current study, there was a detectable effect of the far-red supplementation in terms of stem elongation. However, in lettuce production, this is a detrimental quality characteristic that can be associated with bolting, i.e., an early stage towards flowering, which is usually associated with an increased bitter taste and reduced marketability.

Overall, the photosynthetic rate was significantly higher in green lettuce than in red lettuce and light spectrum had significantly different effects on the physiological traits measured. The general trend was that the light spectra which produced the best overall yield (white light) had significantly lower photosynthetic rates and lower transpiration rates when measured. It has to be noted that final yield is an integration of biomass production, partitioning and respiration over time (45 days) whilst photosynthetic rate is an instantaneous measurement at one point in time. It has been reported that sensitivity of a lettuce plant to lighting spectra is determined by its cultivar's metabolic plasticity [40]. Focusing the lights in the red and blue region (450 nm or 435 nm or a combination of both) had a significant impact on various physiological parameters, including photosynthetic rate, stomatal conductance and transpiration rate, compared with white light. Similar effects were reported in terms of the impact of light spectrum of both basil and lemon balm growth under controlled environment conditions [8–10]. However, in the current study, the significant improvements in the physiological traits did not translate into an improvement in the growth rate of either green or red lettuce. It can only be assumed that the morphological structure and metabolic pathways are species specific. More research is needed for a further understanding of the conflicting findings regarding the photosynthesis parameters and growth traits observed, and it is clear that no inferences on final yield can be drawn from instantaneous photosynthetic activity measurements.

Light spectra had a significant impact on various chemical contents of both lettuce varieties. There was also a significant interaction between light treatment and lettuce type in terms of the impact on vitamin C content. While red/blue450 (1:1) treatment significantly increased the content of vitamin C in red lettuce, it has a very negative impact in green lettuce. White light, however, had a positive impact on the vitamin C content of both green and red lettuce. This finding agrees with other reports that suggest that achieving a positive impact on the content of vitamin C is complex, because metabolism of antioxidant properties in lettuce depends on multicomponent exposure of variety, light quality and several other factors related to the growing conditions [41]. Light spectra also impacted the total phenol content in both lettuce varieties, and focusing the lights in the red/blue region significantly increased the accumulation of phenols. The effect of UV-B was of interest as it was predicted that UV would influence secondary metabolism which would include phenolic content. Interestingly, red lettuce responded positively to supplementation with UV, whilst a negative effect of UV was detected in green lettuce. In agreement with the current findings, it has been reported that blue light and ultraviolet-A (UV-A) significantly increase the total phenolic content in soybean [42]. It was also reported that UV irradiation can improve the accumulation of phenolic compounds with antioxidant properties in lettuce cultivated in controlled environment systems [43].

Light treatment had a significant impact on the total soluble sugar in both green and red lettuces. While white light increases the content of these compounds in red lettuce, it was white spectrum supplemented with UV that promoted the highest level of soluble sugar in green lettuce. This is in agreement with previous studies, which indicated that the soluble sugar content of plants grown under RBW treatment is significantly higher compared to those grown under R/B treatments [44]. The higher sugar content could improve the taste and have a positive impact on palatability. Light spectra also had a significant impact on the total soluble content, and this was significantly higher in red lettuce than in green lettuce.

The positive impact of R/B treatments on the chlorophyll content observed in this study agrees with the findings of Naznin (2019) [31], who reported that chlorophyll A, chlorophyll B, and total chlorophyll content of lettuce, spinach, basil, and pepper was significantly increased under the R/B light treatment. However, this is in contrast to previous reports indicating that the total chlorophyll content of lettuce plants treated with blue and red light was less than that of lettuce plants treated with FL (fluorescent light) (white) [45]. In addition, it has been reported that continuous white LED lighting increased chlorophyll content in lettuce [40]. This could be due to genetic variations and also to differences between the experimental conditions in which the other studies were conducted and those of the current research. The positive impact of far red and green supplementation was observed in red lettuce and not in green lettuce. This could be caused by the different compositions of pigment content. It could be also caused by the genetic variations.

UV supplement did not seem to have an overall great impact on the chemical traits of both lettuce varieties. This could be due to the use of low doses for a short period of time (one week before final harvest). Future research needs to include more duration and intensity treatment combinations in order to unfold the possible role of UV in improving the chemical profile of lettuce.

5. Conclusions

This investigation set out to compare the effects of various light spectra on the growth and physiological parameters of both red and green varieties of lettuce. Whilst light spectra had a significant impact on various traits, it was apparent that the best spectrum was white light. By using R/B or R/Bl/G spectra, some aspects of physiology could be enhanced, especially instantaneous photosynthetic rate leading to quicker growth, but within the parameters of this experiment, this led to stem elongation and reduced marketability. Attention must be focused on the use of R/B light commercial production systems and recognizing that commercial harvest of a potentially smaller product will be achieved earlier than a conventional white light growing system. This could have commercial advantage if the smaller lettuce size was acceptable to the purchaser. It is worth noting that the addition of far-red wavelengths to the spectrum accentuated the stem elongation in both varieties of lettuce, which needs to be used with caution in commercial set-ups.

There was no clear trend of a light spectrum impact on the general chemical profiles measured despite significant differences being measured. UV-B irradiation gave some indication that it could increase phenolic content, but this effect was only clear in red lettuce. More research is still needed to investigate the mechanism of spectrum impact on the chemical profile of lettuce and on other plant species, since this is clearly species-dependent. Moreover, more studies of possible interactions between light intensity and spectra are recommended.

Author Contributions: Conceptualization, A.A.A. (Abdullah A. Alrajhi) and H.Z.R.; methodology, A.S.A.; formal analysis, A.S.A.; investigation, A.S.A. and A.A.I.; resources, A.A.A. (Abdullah A. Alrajhi) and A.A.I.; data curation, A.S.A.; writing—original draft preparation, A.A.A. (Abdullah A. Alrajhi), A.S.A. and H.Z.R.; writing—review and editing, H.Z.R. and M.P.F.; visualization, A.S.A.; supervision, I.M.A. and A.A.A. (Abdullah A. Alsadon); project administration, A.A.A. (Abdullah A. Alrajhi); funding acquisition, A.A.A. (Abdullah A. Alrajhi). All authors have read and agreed to the published version of the manuscript.

Funding: This research was funded by King Abdulaziz City for Science and Technology (grant number: KACST 41-31129).

Acknowledgments: Authors acknowledge funding from King Abdulaziz City for Science and Technology (KACST), and all technical support from the National Centre for Agriculture Technology (NCAT).

Conflicts of Interest: The authors declare no conflict of interest.

References

1. Shamshiri, R.; Kalantari, F.; Ting, K.; Thorp, K.R.; Hameed, I.A.; Weltzien, C.; Ahmad, D.; Shad, Z.M. Advances in greenhouse automation and controlled environment agriculture: A transition to plant factories and urban agriculture. *Int. J. Agric. Biol. Eng.* **2018**, *11*, 1–22. [CrossRef]
2. Benke, K.; Tomkins, B. Future food-production systems: Vertical farming and controlled-environment agriculture. *Sustain. Sci. Pract. Policy* **2017**, *13*, 13–26. [CrossRef]
3. Abidi, F.; Girault, T.; Douillet, O.; Guillemain, G.; Sintes, G.; Laffaire, M.; Ahmed, H.B.; Smiti, S.; Huché-Thélier, L.; Leduc, N. Blue light effects on rose photosynthesis and photomorphogenesis. *Plant Biol.* **2013**, *15*, 67–74. [CrossRef] [PubMed]
4. Chen, X.-L.; Yang, Q.-C.; Song, W.-P.; Wang, L.-C.; Guo, W.-Z.; Xue, X.-Z. Growth and nutritional properties of lettuce affected by different alternating intervals of red and blue LED irradiation. *Sci. Hortic.* **2017**, *223*, 44–52. [CrossRef]
5. Pennisi, G.; Blasioli, S.; Cellini, A.; Maia, L.; Crepaldi, A.; Braschi, I.; Spinelli, F.; Nicola, S.; Fernandez, J.A.; Stanghellini, C. Unraveling the Role of Red: Blue LED Lights on Resource Use Efficiency and Nutritional Properties of Indoor Grown Sweet Basil. *Front. Plant Sci.* **2019**, *10*, 305. [CrossRef]
6. Rihan, H.Z.; Aljafer, N.; Jbara, M.; McCallum, L.; Lengger, S.; Fuller, M.P. The Impact of LED Lighting Spectra in a Plant Factory on the Growth, Physiological Traits and Essential Oil Content of Lemon Balm (*Melissa officinalis*). *Plants* **2022**, *11*, 342. [CrossRef]
7. Rihan, H.Z.; Aldarkazali, M.; Mohamed, S.J.; McMulkin, N.B.; Jbara, M.H.; Fuller, M.P. A Novel New Light Recipe Significantly Increases the Growth and Yield of Sweet Basil (*Ocimum basilicum*) Grown in Plant Factory System. *Agronomy* **2020**, *10*, 934. [CrossRef]
8. Aldarkazali, M.; Rihan, H.Z.; Carne, D.; Fuller, M.P. The growth and development of sweet basil (*Ocimum basilicum*) and bush basil (*Ocimum minimum*) grown under three light regimes in a controlled environment. *Agronomy* **2019**, *9*, 743. [CrossRef]
9. Sims, D.A.; Gamon, J.A. Relationships between leaf pigment content and spectral reflectance across a wide range of species, leaf structures and developmental stages. *Remote Sens. Environ.* **2002**, *81*, 337–354. [CrossRef]
10. Brown, C.S.; Schuerger, A.C.; Sager, J.C. Growth and photomorphogenesis of pepper plants under red light-emitting diodes with supplemental blue or far-red lighting. *J. Am. Soc. Hortic. Sci.* **1995**, *120*, 808–813. [CrossRef] [PubMed]
11. Lin, K.-H.; Huang, M.-Y.; Hsu, M.-H. Morphological and physiological response in green and purple basil plants (*Ocimum basilicum*) under different proportions of red, green, and blue LED lightings. *Sci. Hortic.* **2021**, *275*, 109677. [CrossRef]
12. Kornpointner, C.; Martinez, A.S.; Marinovic, S.; Haselmair-Gosch, C.; Jamnik, P.; Schröder, K.; Löfke, C.; Halbwirth, H. Chemical composition and antioxidant potential of Cannabis sativa L. roots. *Ind. Crops Prod.* **2021**, *165*, 113422. [CrossRef]
13. Zhang, S.; Ma, J.; Zou, H.; Zhang, L.; Li, S.; Wang, Y. The combination of blue and red LED light improves growth and phenolic acid contents in Salvia miltiorrhiza Bunge. *Ind. Crops Prod.* **2020**, *158*, 112959. [CrossRef]
14. SharathKumar, M.; Heuvelink, E.; Marcelis, L.F.M. Vertical Farming: Moving from Genetic to Environmental Modification. *Trends Plant Sci.* **2020**, *25*, 724–727. [CrossRef] [PubMed]
15. Kim, K.; Kook, H.-S.; Jang, Y.-J.; Lee, W.-H.; Kamala-Kannan, S.; Chae, J.-C.; Lee, K.-J. The effect of blue-light-emitting diodes on antioxidant properties and resistance to Botrytis cinerea in tomato. *J. Plant Pathol. Microbiol.* **2013**, *4*, 203.
16. Wu, M.-C.; Hou, C.-Y.; Jiang, C.-M.; Wang, Y.-T.; Wang, C.-Y.; Chen, H.-H.; Chang, H.-M. A novel approach of LED light radiation improves the antioxidant activity of pea seedlings. *Food Chem.* **2007**, *101*, 1753–1758. [CrossRef]
17. Ohashi-Kaneko, K.; Takase, M.; Kon, N.; Fujiwara, K.; Kurata, K. Effect of light quality on growth and vegetable quality in leaf lettuce, spinach and komatsuna. *Environ. Control Biol.* **2007**, *45*, 189–198. [CrossRef]
18. Choi, H.G.; Moon, B.Y.; Kang, N.J. Effects of LED light on the production of strawberry during cultivation in a plastic greenhouse and in a growth chamber. *Sci. Hortic.* **2015**, *189*, 22–31. [CrossRef]
19. Liandong, Q.; Shiqi, L.; Li, X.; Wenyan, Y.; Qingling, L.; Shuqin, H. Effects of light qualities on accumulation of oxalate, tannin and nitrate in spinach. *Trans. Chin. Soc. Agric. Eng.* **2007**, *23*, 201–205.
20. Li, Q.; Kubota, C. Effects of supplemental light quality on growth and phytochemicals of baby leaf lettuce. *Environ. Exp. Bot.* **2009**, *67*, 59–64. [CrossRef]
21. Giliberto, L.; Perrotta, G.; Pallara, P.; Weller, J.L.; Fraser, P.D.; Bramley, P.M.; Fiore, A.; Tavazza, M.; Giuliano, G. Manipulation of the blue light photoreceptor cryptochrome 2 in tomato affects vegetative development, flowering time, and fruit antioxidant content. *Plant Physiol.* **2005**, *137*, 199–208. [CrossRef]
22. Kamal, K.Y.; Khodaeiaminjan, M.; El-Tantawy, A.A.; Moneim, D.A.; Salam, A.A.; Ash-shormillesy, S.M.; Attia, A.; Ali, M.A.; Herranz, R.; El-Esawi, M.A. Evaluation of growth and nutritional value of Brassica microgreens grown under red, blue and green LEDs combinations. *Physiol. Plant.* **2020**, *169*, 625–638. [CrossRef]

23. Kim, H.H.; Wheeler, R.M.; Sager, J.C.; Gains, G.; Naikane, J. Evaluation of Lettuce Growth Using Supplemental Green Light with Red and Blue Light-Emitting Diodes in a Controlled Environment—A Review of Research at Kennedy Space Center. In Proceedings of the V International Symposium on Artificial Lighting in Horticulture 711, Lillehammer, Norway, 21–24 June 2005; pp. 111–120.
24. Legendre, R.; van Iersel, M.W. Supplemental far-red light stimulates lettuce growth: Disentangling morphological and physiological effects. *Plants* **2021**, *10*, 166. [CrossRef] [PubMed]
25. Ferreyra, M.L.F.; Serra, P.; Casati, P. Recent advances on the roles of flavonoids as plant protective molecules after UV and high light exposure. *Physiol. Plant.* **2021**, *173*, 736–749. [CrossRef] [PubMed]
26. Shimizu, H.; Saito, Y.; Nakashima, H.; Miyasaka, J.; Ohdoi, K. Light environment optimization for lettuce growth in plant factory. *IFAC Proc. Vol.* **2011**, *44*, 605–609. [CrossRef]
27. Saito, Y.; Shimizu, H.; Nakashima, H.; Miyasaka, J.; Ohdoi, K. The effect of light quality on growth of lettuce. *IFAC Proc. Vol.* **2010**, *43*, 294–298. [CrossRef]
28. Goins, G.D.; Ruffe, L.M.; Cranston, N.A.; Yorio, N.C.; Wheeler, R.M.; Sager, J.C. *Salad Crop Production under Different Wavelengths of Red Light-Emitting Diodes (LEDs)*; SAE Technical Paper; SAE International: Warrendale, PA, USA, 2001.
29. Wang, J.; Lu, W.; Tong, Y.; Yang, Q. Leaf morphology, photosynthetic performance, chlorophyll fluorescence, stomatal development of lettuce (*Lactuca sativa L.*) exposed to different ratios of red light to blue light. *Front. Plant Sci.* **2016**, *7*, 250. [CrossRef]
30. Wang, X.; Xu, X.; Cui, J. The importance of blue light for leaf area expansion, development of photosynthetic apparatus, and chloroplast ultrastructure of Cucumis sativus grown under weak light. *Photosynthetica* **2015**, *53*, 213–222. [CrossRef]
31. Naznin, M.T.; Lefsrud, M.; Gravel, V.; Azad, M.O.K. Blue Light added with Red LEDs Enhance Growth Characteristics, Pigments Content, and Antioxidant Capacity in Lettuce, Spinach, Kale, Basil, and Sweet Pepper in a Controlled Environment. *Plants* **2019**, *8*, 93. [CrossRef] [PubMed]
32. Tang, Y.; Guo, S.; Ai, W.; Qin, L. *Effects of Red and Blue Light Emitting Diodes (LEDs) on the Growth and Development of Lettuce (var. Youmaicai)*; SAE Technical Paper; SAE International: Warrendale, PA, USA, 2009.
33. Mohamed, S.J.; Rihan, H.Z.; Aljafer, N.; Fuller, M.P. The Impact of Light Spectrum and Intensity on the Growth, Physiology, and Antioxidant Activity of Lettuce (*Lactuca sativa L.*). *Plants* **2021**, *10*, 2162. [CrossRef]
34. Saha, S.; Monroe, A.; Day, M.R. Growth, yield, plant quality and nutrition of basil (*Ocimum basilicum L.)* under soilless agricultural systems. *Ann. Agric. Sci.* **2016**, *61*, 181–186. [CrossRef]
35. Mills, T.; Dunn, B. *LED Grow Lights for Plant Production*; Oklahoma Cooperative Extension Service: Oklahoma City, OK, USA, 2016.
36. Massa, G.D.; Kim, H.-H.; Wheeler, R.M.; Mitchell, C.A. Plant productivity in response to LED lighting. *HortScience* **2008**, *43*, 1951–1956. [CrossRef]
37. Kim, H.-H.; Goins, G.D.; Wheeler, R.M.; Sager, J.C. Green-light supplementation for enhanced lettuce growth under red-and blue-light-emitting diodes. *HortScience* **2004**, *39*, 1617–1622. [CrossRef] [PubMed]
38. Zou, J.; Zhang, Y.; Zhang, Y.; Bian, Z.; Fanourakis, D.; Yang, Q.; Li, T. Morphological and physiological properties of indoor cultivated lettuce in response to additional far-red light. *Sci. Hortic.* **2019**, *257*, 108725. [CrossRef]
39. Viczián, A.; Klose, C.; Ádám, É.; Nagy, F. New insights of red light-induced development. *Plant Cell Environ.* **2017**, *40*, 2457–2468. [CrossRef]
40. Viršilė, A.; Brazaitytė, A.; Vaštakaitė-Kairienė, V.; Miliauskienė, J.; Jankauskienė, J.; Novičkovas, A.; Laužikė, K.; Samuolienė, G. The distinct impact of multi-color LED light on nitrate, amino acid, soluble sugar and organic acid contents in red and green leaf lettuce cultivated in controlled environment. *Food Chem.* **2020**, *310*, 125799. [CrossRef]
41. Samuolienė, G.; Sirtautas, R.; Brazaitytė, A.; Duchovskis, P. LED lighting and seasonality effects antioxidant properties of baby leaf lettuce. *Food Chem.* **2012**, *134*, 1494–1499. [CrossRef]
42. Zhang, X.; Bian, Z.; Li, S.; Chen, X.; Lu, C. Comparative analysis of phenolic compound profiles, antioxidant capacities, and expressions of phenolic biosynthesis-related genes in soybean microgreens grown under different light spectra. *J. Agric. Food Chem.* **2019**, *67*, 13577–13588. [CrossRef]
43. Lee, M.J.; Son, J.E.; Oh, M.M. Growth and phenolic compounds of Lactuca sativa L. grown in a closed-type plant production system with UV-A, -B, or-C lamp. *J. Sci. Food Agric.* **2014**, *94*, 197–204. [CrossRef]
44. Lin, K.-H.; Huang, M.-Y.; Huang, W.-D.; Hsu, M.-H.; Yang, Z.-W.; Yang, C.-M. The effects of red, blue, and white light-emitting diodes on the growth, development, and edible quality of hydroponically grown lettuce (*Lactuca sativa* L. var. capitata). *Sci. Hortic.* **2013**, *150*, 86–91. [CrossRef]
45. Johkan, M.; Shoji, K.; Goto, F.; Hashida, S.-n.; Yoshihara, T. Blue light-emitting diode light irradiation of seedlings improves seedling quality and growth after transplanting in red leaf lettuce. *HortScience* **2010**, *45*, 1809–1814. [CrossRef]

plants

MDPI

Article

Effect of Supplemental Inter-Lighting on Paprika Cultivated in an Unheated Greenhouse in Summer Using Various Light-Emitting Diodes

Yong Beom Kwon [1], Joo Hwan Lee [1], Yoo Han Roh [1], In-Lee Choi [2], Yongduk Kim [3], Jidong Kim [4] and Ho-Min Kang [1,2,*]

1 Interdisciplinary Program in Smart Agriculture, Kangwon National University, Chuncheon 24341, Republic of Korea; nm96727@naver.com (Y.B.K.); ot2581@naver.com (J.H.L.); nuh3722@naver.com (Y.H.R.)
2 Agricultural and Life Science Research Institute, Kangwon National University, Chuncheon 24341, Republic of Korea; cil1012@kangwon.ac.kr
3 Cheorwon Plasma Research Institute, Cheorwon 24062, Republic of Korea; ydkim@cpri.re.kr
4 FutureGreen Co., Ltd., Yongin 17095, Republic of Korea; jidong.kim@futuregreen.co.kr
* Correspondence: hominkang@kangwon.ac.kr; Tel.: +82-33-250-6425; Fax: +82-33-259-5562

Abstract: This study investigated the effects of supplemental inter-lighting on paprika (cv. Nagano RZ) in South Korea in summer using various LED light sources. The following LED inter-lighting treatments were used: QD-IL (blue + wide-red + far-red inter-lighting), CW-IL (cool-white inter-lighting), and B+R-IL (blue + red (1:2) inter-lighting). To investigate the effect of supplemental lighting on each canopy, top-lighting (CW-TL) was also used. Additionally, a control without supplemental lighting was included for comparison. Significant variations were observed in the plant growth indexes 42 days after treatment. The SPAD values and total chlorophyll content in the last period of cultivation were significantly higher than those of the control. In November, the marketable fruit yield was significantly higher than that of the control. QD-IL, CW-IL, and CW-TL resulted in significantly higher values of total soluble solids than the control, and CW-IL resulted in higher values of ascorbic acid content than the control. Regarding the economic analysis, CW-IL resulted in the highest net income rate (12.70%) compared with the control. Therefore, the light sources of CW-IL were assessed as suitable for supplemental lighting due to the highest total soluble solids, ascorbic acid content, and net income rate obtained.

Keywords: paprika (*Capsicum annum* L.); summer-cultivated; supplemental lighting; inter-lighting; top-lighting; LED; economic analysis

Citation: Kwon, Y.B.; Lee, J.H.; Roh, Y.H.; Choi, I.-L.; Kim, Y.; Kim, J.; Kang, H.-M. Effect of Supplemental Inter-Lighting on Paprika Cultivated in an Unheated Greenhouse in Summer Using Various Light-Emitting Diodes. *Plants* **2023**, *12*, 1684. https://doi.org/10.3390/plants12081684

Academic Editors: Hail Rihan and Mick Fuller

Received: 21 March 2023
Revised: 11 April 2023
Accepted: 12 April 2023
Published: 17 April 2023

1. Introduction

Paprika is a high-income crop and a representative fruit vegetable in greenhouses, and as health-focused eating habits are becoming more widespread, the cultivation area of paprika in Republic of Korea is also increasing [1]. Paprika is cultivated in both summer and winter in Korea. Summer cultivation tends to yield about 20% less than winter cultivation because of intense solar radiation, high air temperatures, and humidity [2]. Intense solar radiation causes high temperatures by increasing the radiant heat in a greenhouse. Exposure to high temperatures in flower buds 16 to 18 days before anthesis causes pollen sterility and a reduction in pollen viability, which reduces fruit size and fruit set [3]. Furthermore, due to summer torrential rain, there is a large variation in solar radiation [4]. In addition, Korea has a summer rainy season called the "Changma" due to the monsoon system in East Asia. Recent weather changes have resulted in heavy rainfall, and increased deep convection was evident in August and September [5,6]. In other words, during the summer cultivation of paprika in Korea, high temperatures caused by intense solar radiation can reduce yields.

Additionally, the period of weak sunlight during torrential rain can also be a limitation for paprika cultivation. Previous studies indicated that a decrease in source strength, such as solar radiation, led to a linear increase in the rate of fruit abortion, and it was also reported that a 1% reduction in light resulted in a decrease in the average yield between 0.8 and 1% [7,8]. Shades artificially created to simulate cloudy and rainy days reduced the daily light integral (DLI), which resulted in reduced tomato yields [9]. In addition, fruit vegetables in greenhouses cultivated at high density often lead to excessive mutual shading among plants [10]. Therefore, in order to improve the growth and yield of paprika in greenhouses even in summer, it may be necessary to improve the light environment, not only by shading plants to prevent intense solar radiation but also with supplemental lighting.

Artificial light sources such as a high-pressure sodium (HPS) lamps or light-emitting diodes (LEDs) can be primarily used for supplemental lighting. The choice between HPS lamps and LEDs depends on the application, as HPS lamps have a broader light distribution, which enables a wide area to be covered, and LEDs have a narrower light distribution, with a greater focus on the lighting area [11,12]. However, compared with LEDs, it is difficult to manipulate the spectra of HPS lamps or even to dim them, and they present a great heat emission compared with LEDs [13]. Greenhouse fruit vegetables, such as paprika, mostly use a high-wire training system, i.e., the main stems are supported with a vertical high wire to ensure crop loads [14,15]. Overhead lighting, such as HPS lamps, tends to only focus light on the upper canopy, which can cause mutual shading in the lower canopy and reduce the light reaching the lower parts of high-wire crops [16]. Therefore, in previous studies, supplemental inter-lighting with LEDs has been used in greenhouse fruit vegetables to effectively reduce mutual shading by distributing light to plants [9,16–21].

In summer, supplemental lighting can be restricted due to intense solar radiation and high air temperatures. Previous studies reported that supplemental LED inter-lighting during the summertime increased the stomatal conductance and transpiration rate but did not induce physiological changes in the intra-canopy due to the high DLI, and LED inter-lighting during summertime tomato cultivation resulted in a greater dry matter allocation to leaves and stems than to flowering and fruit development compared with the control [22,23]. However, another study reported that although daytime light in the summer could not improve the yield, nighttime LED inter-lighting had a positive effect on photosynthesis, growth and yields in summer and winter [17]. In the Mediterranean and Jordan Valley, which have a relatively high amount of solar radiation, supplemental LED lighting was used for the purpose of increasing yields, while in other studies, it was hypothesized that supplemental LED lighting could improve the yield, total soluble solids, and ascorbic acid parameters of greenhouse tomato plants even in extremely hot summers when shading is needed [19,20,24]. In paprika cultivation in Korea, supplemental lighting has been mostly applied to winter cultivation, which presents low solar radiation [25–27]. Nevertheless, the hot period is considered in this approach, and supplemental lighting is used during the rainy season, which has a low amount of solar radiation, so an increase in both quantity and quality can also be expected in summer cultivation.

In horticulture, LED fixtures usually combine red, blue, white, and far-red LEDs; in the case of white LEDs, due to their widespread usage, their fraction can increase to more than 60% of the total LEDs [28]. A well-known method for producing white LEDs is the combination of yellow-emitting yttrium aluminum garnet (YAG) phosphor and blue-emitting LED chips, but this kind of white LED lacks the wavelength of strong red, which is mainly used for photosynthesis [29,30]. Recently, quantum dot (QD) materials were applied in white LEDs to increase the strong-red wavelength, and they were applied in blue LEDs to obtain the blue, red, and far-red wavelengths to provide potential application value for agricultural production [29,30].

In conclusion, it is necessary to determine whether the growth and yield increase through LED supplemental lighting is valid in Korean summer paprika cultivation, which is characterized by high temperatures due to strong solar radiation during the cultivation period and a significant decrease in sunlight due to a long rainy season. Furthermore, even though a 2012 study by Jokinen et al. reported that the overall profitability of LED inter-lighting was highly sensitive to yield advantages, product pricing, and installation costs, it also found that when electricity costs and capital costs were combined, LED product prices were still reported to be too high to become profitable at that time [31]. However, many studies still tend to only compare the effects of LED inter-lighting with physiological effects. Therefore, in this study, we used various light sources to select the most suitable option for Korean summer-cultivated paprika, such as LEDs using QD materials, which provide a customized optimal light quality for production, and different lighting positions such as inter-lighting and top-lighting. Additionally, the profitability of supplemental lighting in summer cultivation was investigated through economic analysis. This study aimed to investigate the effect of supplemental lighting with various LEDs, such as those using QD materials, considering the summer climate in South Korea, which is characterized by high temperatures caused by intense solar radiation and long periods of low light due to the rainy season, and examine the effectiveness of supplemental lighting in summer cultivation using economic analysis.

2. Results

2.1. Plant Growth

Regarding the plant growth of paprika, except for the leaf area index (LAI), the other values were not significant 56 days after treatment (DAT) (Table 1). However, plant growth index, plant height, number of nodes, number of leaves, and LAI had different initial values for each paprika plant.

Table 1. Plant growth of paprika after 56 days of supplemental lighting ± SEM (n = 21).

Treatment [z]	Height (cm)	No. of Nodes (ea)	No. of Leaves (ea)	LAI (m^2/m^2)	No. of Flower (ea)	No. of Fruit Set (ea)
QD-IL	155.4 ± 2.0 a [y]	30.86 ± 0.29 a	91.52 ± 1.32 a	4.15 ± 0.14 ab	2.00 ± 0.12 a	9.49 ± 0.47 a
CW-IL	155.5 ± 1.9 a	30.67 ± 0.23 a	90.62 ± 0.89 a	4.21 ± 0.13 ab	1.71 ± 0.14 a	9.39 ± 0.43 a
B+R-IL	158.0 ± 2.1 a	31.14 ± 0.34 a	90.76 ± 0.88 a	4.18 ± 0.11 ab	1.86 ± 0.11 a	9.71 ± 0.40 a
CW-TL	159.3 ± 2.9 a	31.71 ± 0.34 a	92.48 ± 1.41 a	4.48 ± 0.12 a	2.00 ± 0.13 a	9.38 ± 0.49 a
Cont	151.8 ± 2.3 a	30.52 ± 0.36 a	87.90 ± 1.83 a	3.84 ± 0.14 b	1.67 ± 0.15 a	9.33 ± 0.52 a

[z] Treatment included: QD-IL, quantum dot LED inter-lighting; CW-IL, cool-white LED inter-lighting; B+R-IL, blue + red LED inter-lighting; CW-TL, cool-white LED top-lighting; Cont, without supplemental lighting. [y] Means with different letters within column indicate statistically significant differences by Duncan's multiple range test at the 5% level.

The growth indexes were more meaningful in terms of how much they increased during the supplemental lighting treatment compared with the initial value rather than the plant growth state on a specific DAT (Figure 1). Regarding the increases in plant height, number of nodes, and LAI, significant differences were observed between the supplemental-lighting-treated plants and the control at 14, 42, and 28 DAT, respectively, while in the case of the number of leaves, a significant difference was only seen at 42 DAT, and the number of leaves was maintained at the same level in all the experimental groups.

Regarding the number of flowers and number of fruit sets, the number of flowers was maintained at 0~2, and the number of fruit sets was maintained at 5~9 regardless of the supplemental lighting period.

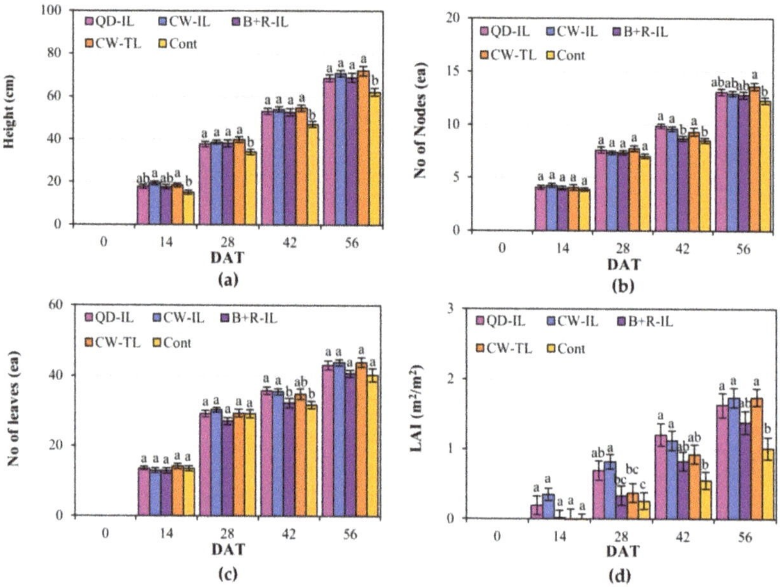

Figure 1. Variance in paprika plant growth during supplemental lighting application. (**a**) Height, (**b**) number of nodes, (**c**) number of leaves, and (**d**) leaf area index (LAI) increased compared with the initial values. Supplemental lighting treatments: QD-IL, QD inter-lighting; CW-IL, cool-white inter-lighting; B+R-IL, blue + red inter-lighting; CW-TL, cool-white top-lighting; Cont, without supplemental lighting. Vertical bars indicate ± SEM (n = 21). Values marked with different letters indicate significant differences according to Duncan's multiple range test at the 5% level.

2.2. *Leaf Characteristics*

The SPAD values were not significant at 56 DAT regardless of the canopy, but the NDVI of the mid-canopy showed a significant difference at 56 DAT between the QD and cool-white inter-lighting treatments compared with the control. In the case of Fv/Fm, there was a significant difference among the treatment groups at 56 DAT, but it was difficult to evaluate whether the effect was due to supplemental lighting (Table 2).

Table 2. Leaf characteristics of paprika after 56 days of supplemental lighting ± SEM (n = 21).

Treatment [z]	SPAD		NDVI		Fv/Fm [y]	
	Top Canopy	Mid Canopy	Top Canopy	Mid Canopy	Top Canopy	Mid Canopy
QD-IL	62.28 ± 0.56 a [x]	67.52 ± 0.54 a	0.632 ± 0.003 a	0.637 ± 0.004 a	0.791 ± 0.014 a	0.784 ± 0.007 c
CW-IL	61.60 ± 0.83 a	65.82 ± 0.74 a	0.636 ± 0.003 a	0.633 ± 0.004 a	0.784 ± 0.010 a	0.790 ± 0.008 c
B+R-IL	62.73 ± 0.64 a	66.56 ± 0.67 a	0.629 ± 0.003 a	0.630 ± 0.004 ab	0.774 ± 0.008 a	0.826 ± 0.007 a
CW-TL	63.93 ± 0.67 a	67.41 ± 0.54 a	0.627 ± 0.003 a	0.626 ± 0.004 ab	0.780 ± 0.006 a	0.812 ± 0.005 ab
Cont	62.57 ± 0.68 a	67.01 ± 0.46 a	0.634 ± 0.003 a	0.620 ± 0.004 b	0.774 ± 0.011 a	0.799 ± 0.008 bc

[z] See Table 1 for details on the treatment included. [y] In the case of Fv/Fm (n = 18). [x] Means with different letters within column indicate statistically significant differences by Duncan's multiple range test at the 5% level.

Regarding the variation in SPAD values after treatment, the top-canopy SPAD values were significantly different from those at 0 DAT. Therefore, the results of the leaf characteristics may have varied for each individual leaf. In addition, regarding the results 28 DAT, which were affected by August and September conditions, which did not meet the standard DLI, the average of the SPAD values of the control were lower than those of the supplemental-lighting-treated plants, even though there were no significant differences. However, the difference between the control group and the supplemental-lighting-treated plants at 42 days was not a significant difference. This means that solar radiation had a greater impact than inter-lighting, which compensated for mutual shading due to the plant height (Figure 2a,b). Moreover, at 56 DAT, the paprika was sufficiently grown, and we pinched out the growing tips so that it was able to continuously receive supplemental lighting in a certain position.

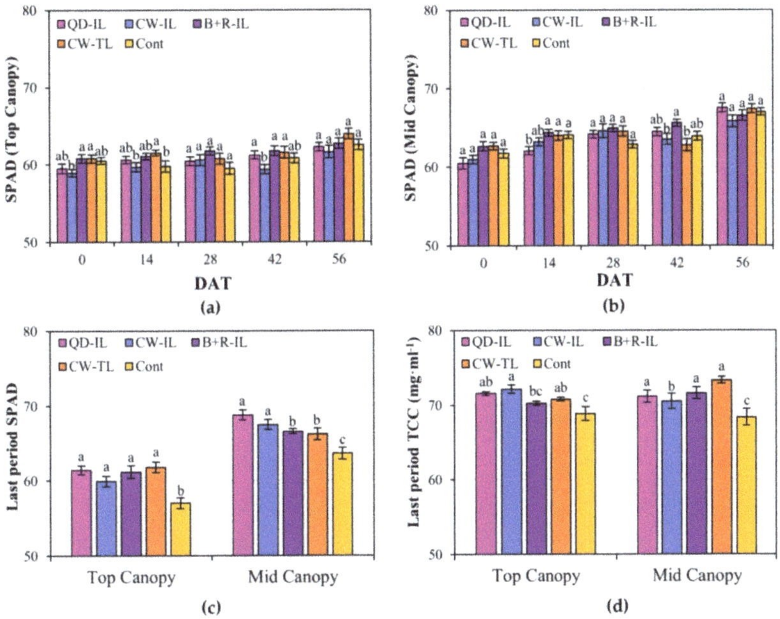

Figure 2. Leaf characteristics, especially chlorophyll content, of paprika plants for each canopy during supplemental lighting application. (**a**,**b**) Variations in SPAD values. (**c**,**d**) SPAD and total chlorophyll content (TCC) of leaves harvested in the last period of cultivation. Supplemental lighting treatments: QD-IL, QD inter-lighting; CW-IL, cool-white inter-lighting; B+R-IL, blue + red inter-lighting; CW-TL, cool-white top-lighting; Cont, without supplemental lighting. Vertical bars indicate ±SEM (n = 21) in (**a**–**c**) and ±SEM (n = 9) in (**d**). Values marked with different letters indicate significant differences according to Duncan's multiple range test at the 5% level.

The results of the SPAD values and total chlorophyll content measured by harvesting leaves at each height in the last period of cultivation were as follows: Regarding the SPAD values, in the case of the top canopy, the supplemental lighting treatment showed a significant difference compared with the control. In the case of the mid-canopy, the QD inter-lighting treatment showed the highest value, and compared with the control, supplemental lighting resulted in significantly higher values (Figure 2c). The total chlorophyll content, in the case of the top canopy, was significantly higher in the plants receiving all the supplemental lighting treatments, except for those receiving the blue + red inter-lighting treatment, than in control plants; in the case of the mid-canopy, this value was significantly higher in the supplemental-lighting-treated plants, except for plants receiving the cool-white inter-lighting treatment, than in the control (Figure 2d). The significant difference in the measured chlorophyll values seems to have occurred due to the sites for the SPAD value measurement and the sampling sites for the total chlorophyll content measurement. However, regarding both values, the supplemental lighting treatment resulted in higher values than the control. Therefore, to induce a significant difference in the chlorophyll content using inter-lighting, conditions of large mutual shading, such as great plant height or high LAI, and continuous supplemental lighting in the same position to illuminate the same leaves are necessary.

2.3. Yield and Fruit Characteristics

In the case of the total fruit set number, there was a difference in September due to the continuous insufficient solar radiation received from August to September. Moreover, in November, the last period of cultivation, the total fruit set number of the supplemental-lighting-treated plants was higher than that of the control plants (Figure 3a).

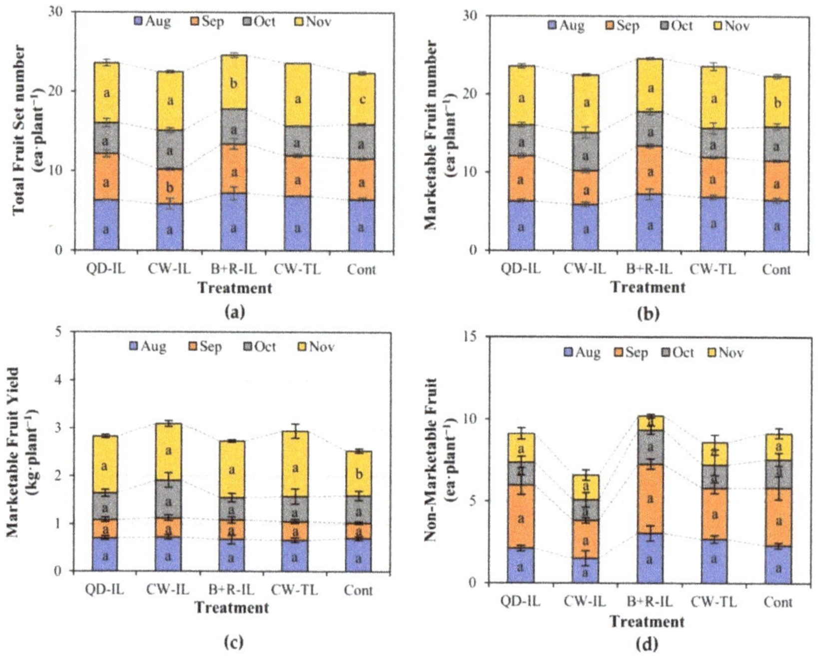

Figure 3. (**a**) Total fruit set number during supplemental lighting application. (**b**) Marketable fruit number during supplemental lighting application. (**c**) Marketable fruit yield during supplemental lighting application. (**d**) Non-marketable fruit number during supplemental lighting application. Marketable fruit means that there were no flaws, such as blossom rot, insect damage, sunburn, or malformation, and the weight was more than 100 g. In November, marketable fruits included green paprika, which was unripe but mature and in equal condition to the marketable variety. Supplemental lighting treatments: QD-IL, QD inter-lighting; CW-IL, cool-white inter-lighting; B+R-IL, blue + red inter-lighting; CW-TL, cool-white top-lighting; Cont, without supplemental lighting. Vertical bars indicate ± SEM (n = 3). Values marked with different letters indicate significant differences according to Duncan's multiple range test at the 5% level.

Marketable fruits refer to fruits excluding those that were dropped or removed due to serious damage and those of less than 100 g in fruit weight. In the case of the marketable fruit number, there were no differences in September, but a significantly higher number was obtained with the supplemental lighting treatment than with the control in November (Figure 3b). Regarding the marketable yield, there was a significant difference between the cool-white top-lighting treatment and the control in November, but no significant differences were found among the other treatments (Figure 3c). The number of non-marketable fruits was the lowest with the cool-white inter-lighting treatment, but this was not significant due to the large standard error between the treatments and blocks (Figure 3d). The marketable yield was the highest with the cool-white inter-lighting treatment, even with the lowest number of fruit sets. Moreover, the lowest number of non-marketable fruits was obtained with the cool-white inter-lighting treatment. Based on these results, there was a difference in the number of fruit sets and the number of marketable fruits obtained with supplemental lighting at harvest in November, at the end of cultivation. Non-marketable fruits can be largely classified into those with blossom rot, insect damage caused by oriental tobacco budworm (*Helicoverpa assaluta*), sunburn, and malformations (Figure 4). Blossom rot occurred in large quantities from August to September, which was a high-temperature period and included a long rainy season. Insect damage occurred in large quantities from September to October. Sunburn and malformations in the paprika fruits rarely occurred. In the case of non-marketable fruit, the degree of occurrence differed for each block and presented large standard errors; thus, the supplemental lighting treatment was not necessarily the cause. However, in the plants receiving the QD inter-lighting treatment and the blue + red inter-lighting treatment with higher blue–light-ratio spectra, there was a tendency of damage due to oriental tobacco budworm. Regarding moths, a higher attractiveness of blue light than white, green, and red light sources was observed [32]. Therefore, it is necessary to check whether the use of LEDs after sunset has any side effects of attracting pests.

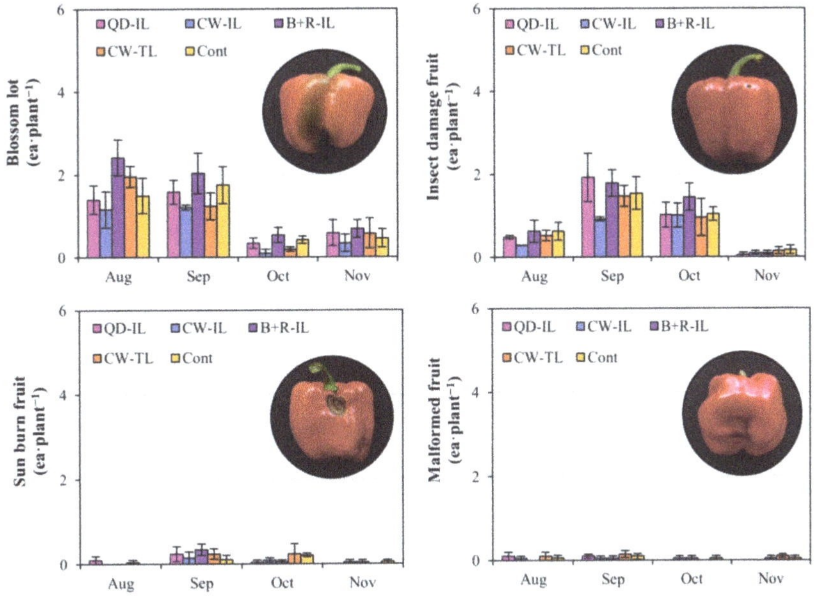

Figure 4. Each type of non-marketable fruit generated during the supplemental lighting period and degree of occurrence. Supplemental lighting treatments: QD-IL, QD inter-lighting; CW-IL, cool-white inter-lighting; B+R-IL, blue + red inter-lighting; CW-TL, cool-white top-lighting; Cont, without supplemental lighting. Vertical bars indicate ±SEM (n = 3). These results were not statistically significant due to the large standard error between each block.

There were no significant differences in the physical characteristics of the fruits due to supplemental lighting (Table 3). Among the internal characteristics of the fruits affected by the number of harvest days, the total soluble solids and ascorbic acid content showed significant differences. In the case of total soluble solids, except for the blue + red inter-lighting treatment, the rest of the supplemental lighting treatments resulted in higher values than the control. In the case of the ascorbic acid content, cool-white inter-lighting resulted in the highest value, showing a significant difference compared with blue + red—inter-lighting and the control (Table 4). Due to the high non-marketable rate caused by the blue + red inter-lighting treatment, compared with those observed with other treatments, we did not observe a significant difference compared with the control. In Appolloni's study (2021), most of the supplemental lighting studies were associated with positive effects on the total soluble solids and ascorbic acid content, but inconsistent results were also found due to other factors [24].

Table 3. Physical characteristics of fruits during the supplemental lighting ± SEM (n = 30).

Treatment [z]	Fruit Weight (g)	Fruit Length (mm)	Fruit Width (mm)	No. of Locules (ea)	Pericarp Thickness (mm)
QD-IL	194.4 ± 5.4 a [y]	90.27 ± 1.55 a	82.52 ± 1.10 a	3.61 ± 0.05 a	6.18 ± 0.11 a
CW-IL	195.4 ± 4.9 a	94.72 ± 1.73 a	82.35 ± 0.94 a	3.49 ± 0.13 a	6.11 ± 0.10 a
B+R-IL	189.0 ± 5.3 a	90.81 ± 1.51 a	80.88 ± 1.07 a	3.61 ± 0.06 a	6.34 ± 0.14 a
CW-TL	192.9 ± 6.0 a	91.22 ± 1.90 a	83.20 ± 1.07 a	3.67 ± 0.08 a	6.05 ± 0.11 a
Cont	197.6 ± 4.8 a	94.58 ± 1.55 a	83.23 ± 1.12 a	3.57 ± 0.07 a	6.21 ± 0.14 a

[z] See Table 1 for details on the treatment included. [y] Means with different letters within column indicate statistically significant differences by Duncan's multiple range test at the 5% level.

Table 4. Fruit characteristics during the supplemental lighting affected by harvest time ± SEM (n = 21).

Treatment [z]	Total Soluble Solids (Brix°)	Ascorbic Acid Contents [y] ($mg{\cdot}100\ g^{-1}$)	Firmness (N)	a Value
QD-IL	7.18 ± 0.05 a [x]	117.9 ± 3.0 ab	37.15 ± 0.67 a	36.28 ± 1.24 a
CW-IL	7.34 ± 0.04 a	121.7 ± 2.3 a	38.52 ± 1.40 a	34.24 ± 0.95 a
B+R-IL	6.87 ± 0.15 b	114.0 ± 1.4 b	35.98 ± 0.97 a	33.37 ± 1.26 a
CW-TL	7.23 ± 0.07 a	117.7 ± 2.8 ab	40.52 ± 1.57 a	34.12 ± 1.02 a
Cont	6.93 ± 0.10 b	111.4 ± 2.2 b	39.41 ± 1.50 a	32.97 ± 0.79 a

[z] See Table 1 for details on the treatment included. [y] In the case of ascorbic acid contents (n = 9). [x] Means with different letters within column indicate statistically significant differences by Duncan's multiple range test at the 5% level.

2.4. Economic Analysis

The net income rate was the highest with the cool-white inter-lighting treatment due to its high marketable yield and the second-lowest total incremental cost (Table 5). The total incremental costs of LED installation per 1000 m^2 for the experimental period were calculated to be USD 1769 for QD inter-lighting, USD 2015 for cool-white inter-lighting, USD 2138 for blue + red inter-lighting, and USD 3137 for cool-white top-lighting. For the same period, the electricity usage costs per 1000 m^2 were calculated to be USD 657 for QD inter-lighting, USD 753 for cool-white inter-lighting, USD 1330 for blue + red inter-lighting, and USD 1137 for cool-white top-lighting. Therefore, almost 70% of the total incremental cost was spent on LED installation.

Table 5. Economic analysis of paprika fruits during the supplemental lighting.

Treatment [z]	Marketable Yield (kg/1000 m^2)	Gross Income [y] (USD/1000 m^2)	Incremental Cost (USD/1000 m^2)			Net Income [v] (USD/1000 m^2)	Net Income Rate [u] (%)
			LED Installation [x]	Electricity Cost [w]	Total		
QD-IL	8970	33,262	1769	657	2425	1370	4.65
CW-IL	9660	35,978	2015	753	2768	3743	12.70
B+R-IL	9120	33,857	2138	1330	3468	921	3.13
CW-TL	9480	34,915	3137	1137	4274	1173	3.98
Cont	7890	29,468	-	-	-	-	0.00

[z] See Table 1 for details on the treatment included. [y] Unit price of red or green paprika per kg × marketable yield per 1000 m^2. [x] (LEDs cost + SMPS cost) × number of light sources per 1000 m^2 ÷ life expectancy of LED ÷ 12 × experiment period. - Life expectancy of LED as 11.42 years, and experiment period as 3.3 months. [w] Electricity consumption of light sources × supplemental lighting time × agricultural electricity cost per kWh × number of light sources per 1000 m^2. - Supplemental lighting time as 6.7 h, and agricultural electricity cost as 40 KRW per kWh. [v] Gross income—(total incremental cost + gross income of Control). [u] Net income ratio = ((gross income of control + net income)/(gross income of control) × 100) − 100.

Since the increase in the gross income due to the increase in the marketable yield obtained with supplemental lighting was confirmed in November, to increase the net income, it is necessary to reduce the total incremental cost using low-cost, high-efficiency LEDs. Additionally, the unit price per kg of green paprika was USD 3.32, which was lower than that of red paprika, whose unit price per kg was USD 3.96. In this experiment, green paprika was sold in November because of concerns about chilling injury. Korean summer cultivation usually involves a cultivation period from June to November. Therefore, it is economically advantageous to distribute the November harvest as red paprika fully ripened in a heated greenhouse.

3. Discussion

In this experiment, the internal environment of the greenhouse reached 30 °C in summer, from June to August. In September, the optimum temperature for growth was reached, and in November, the temperature dropped, so cultivation could not be continued because of concerns about chilling injury. In addition, due to torrential rain, the proportions of days when the internal DLI of the greenhouse was less than 12 mol$\cdot$m$^{-2}\cdot$d^{-1} were 48.4% in August and 60.0% in September. Therefore, in August, the high-temperature period, the occurrence of blossom rot rapidly increased due to the increase in the EC of the medium, and the marketable yield decreased in September, the low-light period.

Inducing significant growth variations in the plant height, number of nodes, and LAI of paprika required supplemental lighting for at least 42 days. The chlorophyll index of the leaf characteristics was measured, and the NDVI showed significant differences for among each of the treatments at 56 DAT. However, SPAD presented both significant and non-significant values during the supplemental lighting period until the pinching out of the growing tips. Therefore, at the end of cultivation, the leaves were harvested to measure their total chlorophyll content together with SPAD. These values were generally higher in the supplemental-lighting-treated plants than in the control plants. In order to increase the chlorophyll content in the leaves with supplemental lighting, at least 56 days were required. In addition, inter-lighting application for a long time under conditions of large mutual shading with an increased plant height or LAI is considered effective for increasing chlorophyll content. In the case of the number of fruit sets, there was a significant difference among the treatments in September, when solar radiation was insufficient, but the difference compared with the control was not large, and there were significant differences in the number of fruit sets, the number of marketable fruits, and the marketable yield compared with November. In terms of the general soluble solids of paprika fruit, the value depends on the period of harvest and the cultivars. Through a comparison of fruit quality among 12 cultivars, it was shown that the range of brix levels was from 6.7 to 9.0 [33]. However, shading has been reported to reduce the soluble solids content of the fruit [9,34]. It has also been reported that higher temperatures during the harvest period result in lower soluble solids contents, and an increase in the soluble solids contents within the fruit is largely due to lower temperatures and assimilated currents [35]. The total soluble solids were generally low due to reduced daily light integral caused by torrential rain and high summer temperatures. However, except for the blue + red inter-lighting treatment, the total soluble solids were significantly higher in the supplemental lighting treatments than in the control. Therefore, summer supplemental lighting can serve as compensation for lower total soluble solids due to an overall harsh environment. However, in the blue + red inter-lighting treatment, the total soluble solids were lower than in the control without supplemental lighting. This may be related to the high levels of non-marketable fruit, such as those with blossom rot, which occurred in the blue + red inter-lighting group around August the most. The general ascorbic acid content of paprika fruit, according to RDA's Korean Food Composition Database, is known to be 91.75 mg per 100 g, but depending on the cultivars, it can range from 55.3 to 189 mg per 100 g [36]. Supplemental inter-lighting can increase the ascorbic acid content of tomato and paprika fruits [17,37]. However, increasing temperatures above 27 degrees can cause an inhibition of ascorbic acid accumulation [38]. Rather than the effect of shading on the ascorbic acid content, it has been reported that increasing the light intensity can increase the ascorbic acid content and stimulate the antioxidant system [9,39]. In this experiment, the ascorbic acid content in the supplemental-lighting-treated group was higher than that in the control, but only the cool white inter-lighting treatment was significantly different from the control. Therefore, the effect of a high temperature on the ascorbic acid content seems to be greater than that on the total soluble solids. The effect of supplemental lighting on the ascorbic acid content was not as significant as that on the total soluble solids, so the difference is expected to be quite small. In the case of blue + red inter-lighting, there was no difference compared to the control. This was consistent with the tendency of total soluble solids, which is related to the high level and rate of non-marketable fruit, such as those with blossom rot, in August. In addition, the degree of fruit maturity can affect the total soluble solids and firmness, and several reports have shown that LED light can affect the harvest time [40]. Therefore, fruits that were harvested at the same number of days after full bloom were compared. The internal quality, such as the total soluble solids and ascorbic acid contents, showed a significant difference in the supplemental LED lighting treatment. However, there was no significant difference in the firmness, which is related to the cellular texture of the paprika pericarp. The Hunter a value showed no significant difference between the treatments. According to Kim's research, in the case of paprika fruit color, unlike tomatoes, paprika has an irregular skin surface color; thus, there was no significant difference in color, but the individual carotenoid content was significantly different according to the supplemental lighting [37]. Overall, the effects of supplemental lighting were observed in terms of the growth, yield, and fruit characteristics. However, it was difficult to determine the effects of the supplementing light in a short period of time during the rainy season, from August to September. Inter-lighting is known as an effective supplemental lighting method to eliminate mutual shading and illuminate lower canopies [16]. In this experiment, the observed effects relative to August~September, i.e., the low-light period, are thought to be due to the fact that the plant height or LAI was still too low for mutual shading to occur. In addition, it is thought that the effect of the supplemental lighting was relatively low due to the stress caused by the high temperature, which was above the optimum growth temperature, along with the low-light period. It has been reported that shade-induced stress and high-temperature stress have the same susceptible effect on paprika plants, act with the same process, leading to fruit and flower abscission, and may also act on the assimilates available for flower and fruit development [7]. In this experiment, shade stress was compensated for by using supplemental lighting, but the temperature increase inside the greenhouse could not be suppressed only by shade cloth. Therefore, non-marketable fruits, such as fruits affected by blossom rot, and the rate of fruit dropping were high, even if insect-damaged fruits were excluded. In conclusion, it is considered that it is more effective to perform long-term supplemental lighting in a greenhouse, which is a more controlled environment where year-round cultivation is possible, than using supplemental lighting for short-term cultivation.

Among the LEDs used in this experiment, the QDs had a wide range of red and far-red spectra along with blue spectra. A previous study reported that the effects of far-red supplemental lighting on greenhouse tomatoes in the off season were a high total soluble solids and a high dry matter rate, and the quality was improved enough to be recognized by consumer panelists [41]. In contrast, there was an increase in the yield when supplemental lighting was used in Mediterranean tomato cultivation, but there were no effects of adding far-red lighting at that latitude; in the cultivation of paprika with far-red supplemental lighting, the yield increased but the carotenoid content decreased, indicating an antagonistic relationship [37,42]. On the other hand, there was also a report indicating that the use of far-red supplemental lighting in paprika resulted in dry matter being distributed in the branches and stems and in a reduced number of fruit sets [43]. According to these reports, the effect of far-red supplemental lighting is influenced by various factors, such as the cultivation location and the ratio of far-red light. In this experiment, the treatment with QDs with far-red supplemental lighting resulted in higher values of growth, marketable yield, and total soluble solids than the control, and so did the cool-white treatment. However, the quality of paprika, such as the level of carotenoid pigments in fruits determined according to the wavelength, requires additional confirmation. The cool-white LEDs used in the two treatments of inter-lighting and top-lighting were characterized by a wide green wavelength with a high color temperature. During the experiment, supplemental lighting was applied around sunrise and sunset. At those times, the ratios of the spectrum to the light intensity of the red and green wavelengths were found to be lower than those of the blue wavelengths and those in the daytime. At twilight, the daylight spectrum changes very rapidly and can trigger a strong response in plants [44]. Therefore, it is expected that this effect can be obtained using supplemental lighting with an adequate light spectrum during the sunrise and sunset periods. Green light is known to indicate that the loss of absorptance efficiency due to the sieve effect is small, and it is also known to indicate that the increases in absorptance efficiency are due to the détour effect [45]. In addition, green light drives photosynthesis more effectively than red light in white light at high PPFDs even though red light is greater at low PPFDs [45].

In a research study on paprika treated with supplemental LED inter-lighting in Jordan Valley by Joshi et al. (2019), inter-lighting using cool-white and RGB light resulted in the highest yield, and it was reported that the green spectral component in cool-white LEDs could be advantageous for inter-lighting [19]. In the case of the two cool-white supplemental lighting treatments used in this experiment, the growth, marketable yield, and fruit characteristics were higher than those of the control, but there were no clear differences between the QDs or blue + red lighting and other wavelengths. However, the cool-white inter-lighting induced the highest marketable yield, and the non-marketable fruit rate was relatively low. Therefore, it is necessary to check whether there are side effects, such as an attraction of specific pests (ex. *Helicoverpa assulta*), depending on the spectrum of LEDs when using supplemental LED lighting at sunset.

As a result of the economic analysis, all the supplemental lighting treatments resulted in a higher net income than the control. Among the supplemental lighting treatments, cool-white inter-lighting resulted in the highest net income rate due to the appropriate total incremental costs and the highest marketable yield. It is important to increase the marketable yield to increase the net income rate compared with the control, but it is also important to reduce the total incremental cost through the use of inexpensive and efficient light sources. QD inter-lighting had a lower marketable yield than the two cool-white treatments, but it had the lowest installation costs and electricity costs. Therefore, it is necessary to compare whether it is possible to reduce the costs when manufacturing cool-white wavelength lighting using QD inter-lighting. In particular, top-lighting was very effective in some indexes, such as the November marketable yield and top-canopy leaf SPAD value, but in terms of net income, it resulted in lower values than the QD inter-lighting or cool-white inter-lighting. It seems that more LEDs would be needed for the top-lighting to produce the same intensity as inter-lighting at a certain distance from the paprika plants; as a result, the incremental cost increased, and the net income decreased. Blue + red inter-lighting resulted in the lowest net income among the supplemental lighting treatments. Although non-marketable fruits occurred the most with this treatment, the reason for the low net income seemed to be that the electricity cost out of the total incremental cost was the highest compared with the rest of the light sources.

4. Materials and Methods

4.1. Plants and Growth Conditions

This study was conducted from 23 May 2022 to 9 November 2022 in an unheated, multi-span plastic greenhouse (width of 13 m, length of 28 m, and eave height of 2.5 m) at Kangwon National University, located in Chuncheon, Gangwon-do (37°52′18.6″ N, 127°44′45.9″ E). The plant material was paprika (*Capsicum annuum* L. cv. Nagano RZ), which is commonly cultivated due to the smallest variation in fruit set between fruit groups, and its high yield and long storage life was used to assess the effect of supplemental lighting on summer cultivation [33,46,47]. The paprika was raised for 4 weeks at Gangwon-do Agricultural

Research & Extensions Services and then transferred to a greenhouse. Three paprika seedlings were transplanted at 160 × 33 cm intervals on 100 × 20 × 10 cm Coir slabs (BIOGROW DUO; Biogrow, Mas de la Fabrègue, France). Before transplanting, the Coir slabs were fully hydrated with paprika standard nutrient solution with an EC of 2.0 $dS \cdot m^{-1}$. Irrigation was determined by considering the weather conditions, the growth stage of paprika, the amount of drainage, and the drainage EC. The irrigation EC was between 2.0 and 3.0 $ds \cdot m^{-1}$, and irrigation was performed using 100~150 mL at a time. Irrigation lasted from 1.5~2 h after sunrise up to 3~4 h before sunset. On sunny days, plants were irrigated with irrigation drippers 8~10 times a day. On cloudy days, plants were only irrigated before noon. Paprika was cultivated with twin-head systems that trained the main stem into a "V" shape, branching from 4~5 nodes [15]. In order to maintain the vegetative stage in early cultivation, flowers up to 3 nodes above the branching point were thinned out. To ensure that each node had an adequate number of leaves, one leaf on the main stem and another on each of the lateral stems were kept. When the height of a paprika plant exceeded the greenhouse eave height, we pinched out the growing tips of the shoots.

During cultivation, the internal greenhouse environmental data of air temperature, relative humidity, and integrated solar radiation were monitored in real time with environmental sensors (ioCrops Clima; ioCrops, Seoul, Republic of Korea) placed at the center of the greenhouse. The external greenhouse environmental data were obtained using an automated surface-observing system (ASOS) provided by Korea Meteorological Administration (KMA) (Figure 5a,b). The annual mean value of photosynthetically active radiation (PAR) is 45% of solar radiance, which can be changed by the intensity and photoperiod, and can also be estimated with solar radiance measurements [48,49]. These estimations are particular to the studied region, but the principles can be applied generally [48]. In this study, the estimation of the DLI was based on a report from Korea by Lee et al. (2002) [50] (Figure 5c). The estimated DLI values were compared based on 12 $mol \cdot m^{-2} \cdot d^{-1}$, which is known as the minimum DLI required for the paprika production cycle [51]. Regarding the solar spectrum inside the greenhouse, a handheld spectrometer (MK350S; UPRtek, Zhunan, Taiwan) was used to measure the spectrum after sunrise (07:00~08:00), around noon (13:00~14:00), and before sunset (18:00~19:00) at 4-week intervals at the center of the greenhouse, and the average values were used to indicate the wavelength according to the relative intensity (Figure 5d). A 55% aluminum screen was installed in the greenhouse for shading. If the temperature in the greenhouse increased to 30 °C and the solar radiation intensity exceeded 300 W/m^2, causing the paprika leaves to start wilting, we added the shade cloth. When the solar radiation intensity dropped below 100 W/m^2, we removed the shade cloth.

4.2. LED Fixtures and Supplemental Lighting

Inter-lighting LED light sources were created using 120 cm bar-type LED fixtures with a 40 W power each. We used QDs that included blue- and a wide range of red- and far-red-wavelength LEDs (Cheorwon Plasma Research Institute, Gangwon-do, Republic of Korea), cool-white-wavelength LEDs with a color temperature of 5700 K (HT400-5700; BISSOL LED, Seoul, Republic of Korea), and blue+red-wavelength LEDs with a ratio of blue to red of 1:2 (HT402-1; BISSOL LED, Seoul, Republic of Korea) (Figure 6a). To ensure inter-lighting, we employed a flat hanger bracket to affix two LEDs in opposite directions without an overlap, which illuminated the inner canopy. In addition, they were designed to be height-adjustable with wire so that they could be suspended in the greenhouse. The light intensity was adjusted to around 145 ± 5 $\mu mol \cdot m^{-2} \cdot s^{-1}$ at a distance of 10 cm from the light sources. To determine the effects of supplemental lighting on each canopy, top-lighting LEDs for overhead lighting were produced using cool-white LEDs.

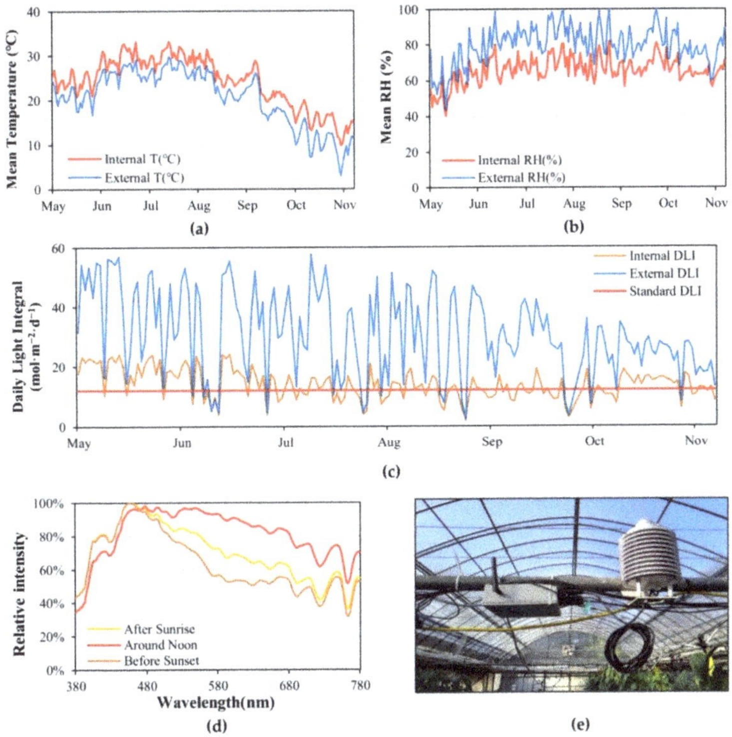

Figure 5. Changes in the greenhouse environment during the experimental period. (**a**) Mean temperature (°C). (**b**) Mean relative humidity (%). (**c**) Daily light integral (mol·m^{-2}·d^{-1}), where the standard DLI was based on the minimum DLI required for paprika. (**d**) Changes in solar spectrum over time: after sunrise (07:00~08:00), around noon (13:00~14:00), and before sunset (18:00~19:00). (**e**) Image of sensor measuring the greenhouse environment (iocrops Clima).

To secure the same level of light intensity as inter-lighting, we arranged three bar-type LEDs without overlapping because the top-lighting structure was at a certain distance from the upper canopy of the paprika. We employed a 300 × 170 mm plastic panel, of which only two ends were bound together to minimize the shading caused by the light sources. In addition, the LEDs were designed to be height-adjustable with wire so that they could be suspended in the greenhouse. Top-lighting LEDs were made to be dimmable to control the light intensity when the height of the LEDs could not be increased because the greenhouse eave height restricted it. The spectrum of each LED fixture used was the average value measured three times during the experimental period. To show the ratio of the wavelengths for each LED fixture, 100% stacked bar graphs of PPF-UV, PPF-B, PPF-G, PPF-R, and PPF-NIR were used (Figure 6b). LEDs for supplemental lighting were installed at 2/3 of the plants based on the slabs between the paprika stems when the paprika plants had grown enough, on 23 July, to illuminate the intra-canopy (Figure 7). Top-lighting was installed 30 cm above the fully developed upper canopy leaves below the growing point, and the light intensity was the same as that of the inter-lighting. Supplemental lighting was applied during 16 h photoperiods, i.e., 04:00~20:00. However, when the greenhouse temperature and solar radiation exceeded 30 °C and 100 W/m^2, respectively, supplemental lighting was discontinued during the day. Every four weeks, the inter-lighting LED position was adjusted to illuminate the intra-canopy at 2/3 of the plants based on the slabs according to the growth of the paprika plants. Similarly, the top-lighting LED position was also adjusted every four weeks. When the top-lighting LEDs reached the greenhouse eave height and we could not adjust the height of the LEDs, we dimmed the light intensity in alignment with the inter-lighting.

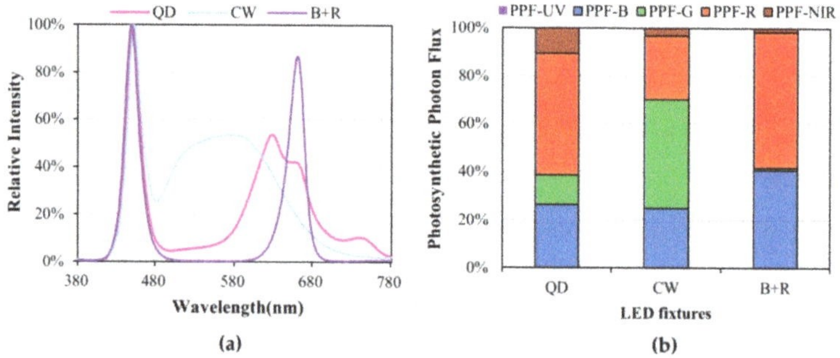

Figure 6. (**a**) Spectrum relative intensity of LED fixtures used in the experiment. QD, quantum dot; CW, cool-white; B+R, blue + red. (**b**) PPF-UV (380~399 nm), PPF-B (400~499 nm), PPF-G (500~599 nm), PPF-R (600~699 nm), and PPF-NIR (700–780 nm) 100% stacked bar graph for the LED fixtures used in the experiment.

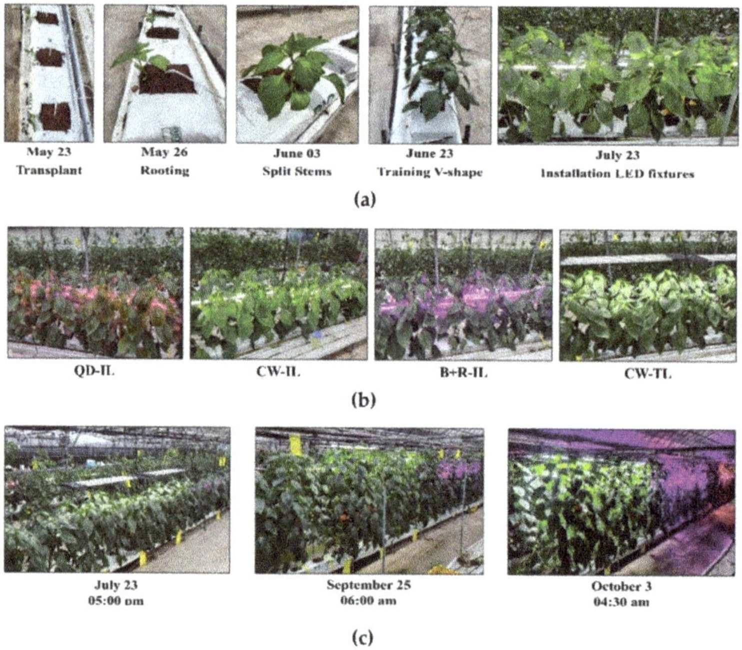

Figure 7. Representative images taken during cultivation. (**a**) Cultivation overview and LED installation. (**b**) Supplemental lighting images for each of the LED treatments: QD-IL, QD inter-lighting; CW-IL, cool-white inter-lighting; B+R-IL, blue + red inter-lighting; CW-TL, cool-white top-lighting. (**c**) View of cultivation sites by period during supplemental lighting application.

4.3. *Measurements*

The height, number of nodes, LAI, number of flowers, and number of fruits were measured to determine the effect of the supplemental lighting on the paprika growth. The leaf area (LA) of one fully developed leaf each from the top canopy and the mid-canopy was estimated by referring to Lee's research (2018). The LA of each leaf was calculated as the average of the leaf area of one leaf per paprika plant; then, the LAI was estimated using the formula obtained by modifying Jang's method (2018) [52,53]:

$$\text{LAI} = \overline{\text{LA}}\left(\text{m}^2\right) \times \frac{\text{No.of leaves}\left(\text{plant}^{-1}\right)}{2} \times \text{Stem density}\left(\text{stems}\cdot\text{m}^{-2}\right) \quad (1)$$

Measurements were repeated 7 times at 14-day intervals after the supplemental lighting application until the pinching out of the growing tips for each block.

The SPAD value, spectral reflectance parameters, and maximum quantum yield of PS II (Fv/Fm) of the paprika leaves were measured to determine the effect of the supplemental lighting on the paprika leaf characteristics for each canopy leaf. Regarding leaf characteristics, the fully developed leaves below the growing tips were referred to as top-canopy leaves, whereas the leaves illuminated by inter-lighting LEDs were referred to as mid-canopy leaves. The SPAD values for each plant were measured three times, and the average values were used; measurements were conducted on fully developed healthy leaves using a chlorophyll meter (SPAD-502; Minolta Camera Co. Ltd., Tokyo, Japan). The spectral reflectance parameters of the leaves were measured using fully developed healthy leaves with a portable spectrophotometer (Polypen RP 410 UVIS; Photon System Instruments, Drásov, Czech Republic). Among the measured values, the NDVI, which is known to be sensitive to the chlorophyll content, was used [54]. The leaf measurements of the SPAD values and spectral reflectance parameters were repeated 7 times at 14-day intervals after supplemental lighting application until the pinching out of the growing tips for each block. The maximum quantum yield of PS II according to chlorophyll fluorescence was measured 20 min after dark adaptation using a PAM fluorometer (JUNIOR PAM; Heinz Walz GmbH, Effeltrich, Germany). Fv/Fm was measured at the same time in the morning on a sunny day, and 6 repetitions were performed for each block on the 28th, 42nd, and 56th days after the supplemental lighting treatment. At the end of cultivation, the leaves of the top canopy and mid-canopy were harvested for the determination of the SPAD values, and the total chlorophyll content was measured. For measurements, 1 g of finely chopped fresh paprika leaves was dissolved in 10 mL of methanol and extracted at 4 °C for 48 h. Then, the absorbance was measured at 642.5 nm and 660 nm using a UV–vis spectrophotometer (BioMate 3S UV-Vis; Thermo Fisher Scientific, Boston, MA, USA). The measuring method referred to was that reported in Yoon's research study (2018) [55,56]. The number of measurements for the total chlorophyll contents was 9, 3 measurements for each block.

Paprika fruits more than 60% ripe were harvested at intervals of about 7 days. The harvested fruits were transferred to the laboratory to check their weight and marketability. For the yield per plant and the number of marketable fruits, all the fruits were harvested and calculated separately by month for each block. Marketable fruits were classified as those not affected by blossom rot, insect damage, sunburn, or malformation and those weighing more than 100 g. In the case of November, due to concerns about chilling injury from the drop in temperature in the greenhouse, all fruits, even unripe green peppers, were harvested on November 9th to check the yield per plant, and marketability was also investigated. During cultivation, instances of significant harm resulting from blossom rot and insect damage during fruit development were recorded, and the affected fruits were eliminated to confirm the number of fruit sets and the number of non-marketable fruits. The number of fruit sets and the number of non-marketable fruits were calculated separately by month for each block. The physical characteristics of the fruits such as the yield per plant, length, width, number of locules, and pericarp thickness were investigated. For the physical characteristics, 7 average-sized fruits from a single harvest were selected to obtain an average value representative of that harvest. The average value of each individual harvest was then used as a sample to represent the average value of the entire harvest period. The harvest dates that were not sufficient to be considered representative of a single harvest were excluded from the calculation. Total soluble solids, ascorbic acid content, firmness, and color may change depending on the harvesting period. Therefore, a total of 18 flowers in full bloom, 6 flowers for each block, were tagged three times on 28 July, 15 August, and 30 August, respectively. Then, fruits that were more than 80% ripe were harvested at the same time for comparison. After extracting juice from the fruit, the total soluble solids were measured with a Brix–acidity meter (PAL-BX ACID 1; Atago Co. Ltd., Tokyo, Japan). The number of measurements for the total soluble solids was 21, 7 measurements for each of the 3 harvest seasons. The ascorbic acid content was determined based on the reduction of yellow molybdophosphoric acid to phosphomolybdenum blue. A reflectometer (RQflex plus; Merck, Darmstadt, Germany) and an ascorbic acid tester (Ascorbic Acid Test; Supelco, Bellefonte, PA, USA) were used for the measurements. To obtain the samples for measurement, we put 2 g of fresh fruit pulp in a tube, which we filled up to 20 mL with distilled water, and homogenized it. After centrifugation at 15,000 rpm at 4 °C for 15 min (Mega 17R; Hanil, Seoul, Republic of Korea), the supernatant was obtained and filtered with a 0.45 µm syringe filter (Minisart® Syringe Filters; Sartorius, Göttingen, Germany). Referring to Ribes-Moya's study (2018), the content of ascorbic acid per 100 g of fruit was measured [57]. The number of measurements for ascorbic acid content was 9, 3 measurements for each of the 3 harvest seasons. For fruit firmness, the paprika pericarp was sliced lengthwise into flat pieces measuring about 30 × 50 mm. The load required to penetrate the pericarp with an 8mm diameter stainless-steel probe was measured using a rheometer (Compac-100II; Sun Scientific Co., Ltd., Tokyo, Japan), and the result was expressed in N (Newton). The number of measurements for firmness was 21, 7 measurements for each of the 3 harvest seasons. Regarding the color, the Hunter a value, which indicates redness, was measured using a color reader (CR-20; Konica Minolta, Tokyo, Japan). The number of measurements for the Hunter a value was 21, 7 measurements for each of the 3 harvest seasons.

Economic analysis. In the economic analysis, we assumed that supplemental lighting was applied for an average of 6.7 h per day for 100 days (3.3 months). Using the yield per plant obtained in this experiment, we assumed a stem density of 6.0 stems·m^{-2} of the cultivation area for 1000 m^2, and the yield of each supplemental lighting treatment compared with the control without supplemental lighting was investigated. The analysis method was carried out with the calculation reported in Hwang's research study (2022) [58].

$$\text{Total incremental cost} = \text{LED installation cost} + \text{Electricity cost} \tag{2}$$

$$\text{Net income} = \text{Gross income} - [\text{Total incremental cost} + \text{Gross income of Control}] \tag{3}$$

$$\text{Net income ratio} = \left[\frac{\text{Gross income of control} + \text{Net income}}{\text{Gross income of control}} \times 100\right] - 100 \tag{4}$$

The LED installation cost share in the total incremental cost was calculated using the unit price at the time of purchase of the LEDs with SMPS, while the electricity cost was determined using a rate of KRW 40 per kWh. The annual cost of paprika per kg used in the calculation of gross income was provided by the Korea Agro-Fisheries & Food Trade Corporation comprehensive agricultural distribution information system (www.nongnet.or.kr). The currency was converted from KRW to USD based on the average exchange rate during the cultivation period (KRW 1342.07 = USD 1).

4.4. Experimental Design and Statistical Analysis

The treatments were arranged in a randomized complete block design (RCBD) with three replicates of each treatment to minimize the effect of other environmental factors, and each treatment was applied to 7 plants. The treatments with supplemental inter-lighting were named "QD-IL", "CW-IL", and "B+R-IL", and the treatment with supplemental top-lighting was named "CW-TL". Additionally, a control treatment ("Cont") without supplemental lighting was included for comparison. Buffer plants were used between two adjacent treatments to avoid light interactions. Statistical analysis was performed using analysis of variance (ANOVA) with DMRT (Duncan's multiple range test) at the 5% significance level. Descriptive data were tested in IBM SPSS statistics, version 26.0 (IBM Corp., Chicago, IL, USA).

5. Conclusions

This study investigated the effect of supplemental LED inter-lighting on summer-cultivated paprika in Korea. Significant differences were observed among the supplemental lighting treatments in terms of the paprika growth variation, the number of fruits, and the marketable yield, which were generally higher than those of the control plants, which were not treated with supplemental lighting. According to the results of the economic analysis, in the case of the net income rate, compared with the control, the cool-white inter-lighting treatment, which resulted in the lowest number of non-marketable fruits, resulted in the highest net income rate (12.70%), and the rest of the supplemental lighting treatments resulted in net income rates of 3~4%. Thus, summer supplemental lighting is effective. However, for more efficient supplemental lighting, it is expected that it would be more efficient to perform long-term supplemental lighting in a greenhouse where the eave height is high and cultivation is possible throughout the year. In addition, research should continue on lower-cost and more effective light sources, such as cool-white LEDs manufactured with QD materials. This would aim to lower installation and electricity costs while increasing the marketable yield. For future work, it would also be interesting to compare the effect of supplemental lighting treatments among various cultivars of paprika, as each cultivar has different pigments.

Author Contributions: Conceptualization, Y.B.K., I.-L.C. and H.-M.K.; methodology, Y.B.K., I.-L.C. and H.-M.K.; formal analysis, Y.B.K., J.H.L. and I.-L.C.; investigation, Y.B.K., J.H.L. and Y.H.R.; resources, Y.K., J.K. and H.-M.K.; data curation, Y.B.K., J.H.L., Y.H.R., I.-L.C. and H.-M.K.; writing—original draft preparation, Y.B.K.; writing—review and editing, Y.B.K., I.-L.C. and H.-M.K.; visualization, Y.B.K.; supervision, I.-L.C. and H.-M.K.; project administration, Y.B.K.; funding acquisition, I.-L.C. and H.-M.K. All authors have read and agreed to the published version of the manuscript.

Funding: This work was supported by the Korea Institute of Planning and Evaluation for Technology in Food, Agriculture, and Forestry (IPET) through the Technology Commercialization Support Program, funded by the Ministry of Agriculture, Food, and Rural Affairs (MAFRA) (122056-3), and the Republic of Korea and the Basic Science Research Program through the National Research Foundation of Korea (NRF) funded by the Ministry of Education (NRF-2021R1A6A1A03044242).

Data Availability Statement: Data is contained within the article.

Acknowledgments: The authors express special thanks to Cheorwon Plasma Research Institute for the supporting QD-LED fixtures for this experiment.

Conflicts of Interest: The authors declare no conflict of interest.

References

1. KATI. *The State of Sweet Pepper Industry in Korea*; Korea Agro-Fisheries Trade Corporation: Seoul, Republic of Korea, 2022. Available online: https://www.kati.net/product/basisInfo.do?lcdCode=MD147 (accessed on 27 January 2023).
2. Cho, I.-H.; Lee, W.-M.; Kwan, K.-B.; Woo, Y.-H.; Lee, K.-H. Stable production technique of paprika (*Capsicum annuum* L.) by hydrogen peroxide treatment at summer. *J. Bio-Environ. Control* **2009**, *18*, 297–301.
3. Erickson, A.; Markhart, A. Flower developmental stage and organ sensitivity of bell pepper (*Capsicum annuum* L.) to elevated temperature. *Plant Cell Environ.* **2002**, *25*, 123–130. [CrossRef]
4. Shin, J.H.; Son, J.E. Irrigation criteria based on estimated transpiration and seasonal light environmental condition for greenhouse cultivation of paprika. *J. Bio-Environ. Control* **2015**, *24*, 1–7. [CrossRef]
5. Yihui, D.; Chan, J.C. The East Asian summer monsoon: An overview. *Meteorol. Atmos. Phys.* **2005**, *89*, 117–142. [CrossRef]
6. Song, H.-J. Long-term variations of cloud top patterns associated with heavy rainfall over the Korean peninsula. *J. Hydrol. Reg. Stud.* **2023**, *46*, 101337. [CrossRef]
7. Marcelis, L.; Heuvelink, E.; Baan Hofman-Eijer, L.; Den Bakker, J.; Xue, L. Flower and fruit abortion in sweet pepper in relation to source and sink strength. *J. Exp. Bot.* **2004**, *55*, 2261–2268. [CrossRef]
8. Marcelis, L.; Broekhuijsen, A.; Meinen, E.; Nijs, E.; Raaphorst, M. Quantification of the growth response to light quantity of greenhouse grown crops. *Acta Hortic.* **2006**, *711*, 97–104. [CrossRef]

9. Tewolde, F.T.; Shiina, K.; Maruo, T.; Takagaki, M.; Kozai, T.; Yamori, W. Supplemental LED inter-lighting compensates for a shortage of light for plant growth and yield under the lack of sunshine. *PLoS ONE* **2018**, *13*, e0206592. [CrossRef]
10. Jiang, C.; Johkan, M.; Maruo, T.; Hohjo, M.; Tsukagoshi, S.; Ebihara, M.; Nakaminami, A. Effect of supplemental far-red light with blue and red LED lamps on leaf photosynthesis, stomatal regulation and plant development of protected cultivated tomato. *Acta Hortic.* **2018**, *1227*, 533–540. [CrossRef]
11. Hernandez, E.; Timmons, M.B.; Mattson, N.S. Quality, Yield, and Biomass Efficacy of Several Hydroponic Lettuce (*Lactuca sativa* L.) Cultivars in Response to High Pressure Sodium Lights or Light Emitting Diodes for Greenhouse Supplemental Lighting. *Horticulturae* **2020**, *6*, 7. [CrossRef]
12. Nelson, J.A.; Bugbee, B. Economic analysis of greenhouse lighting: Light emitting diodes vs. high intensity discharge fixtures. *PLoS ONE* **2014**, *9*, e99010. [CrossRef] [PubMed]
13. Ouzounis, T.; Rosenqvist, E.; Ottosen, C.-O. Spectral effects of artificial light on plant physiology and secondary metabolism: A review. *HortScience* **2015**, *50*, 1128–1135. [CrossRef]
14. Salas, M.; Urrestarazu, M.; Castillo, E. Effect of cultural practices on a sweet pepper crop in a mild winter climate. *Acta Hortic.* **2003**, *614*, 301–306. [CrossRef]
15. Hao, X.; Wen, G.; Papadopoulos, A.P.; Khosla, S. A twin-head "V" high-wire greenhouse cucumber production system for reducing crop start-up costs. *HortTechnology* **2010**, *20*, 963–970. [CrossRef]
16. Gomez, C.; Morrow, R.C.; Bourget, C.M.; Massa, G.D.; Mitchell, C.A. Comparison of intracanopy light-emitting diode towers and overhead high-pressure sodium lamps for supplemental lighting of greenhouse-grown tomatoes. *HortTechnology* **2013**, *23*, 93–98. [CrossRef]
17. Tewolde, F.T.; Lu, N.; Shiina, K.; Maruo, T.; Takagaki, M.; Kozai, T.; Yamori, W. Nighttime supplemental LED inter-lighting improves growth and yield of single-truss tomatoes by enhancing photosynthesis in both winter and summer. *Front. Plant Sci.* **2016**, *7*, 448. [CrossRef]
18. Guo, X.; Hao, X.; Khosla, S.; Kumar, K.; Cao, R.; Bennett, N. Effect of LED interlighting combined with overhead HPS light on fruit yield and quality of year-round sweet pepper in commercial greenhouse. *Acta Hortic.* **2016**, *1134*, 71–78. [CrossRef]
19. Joshi, N.C.; Ratner, K.; Eidelman, O.; Bednarczyk, D.; Zur, N.; Many, Y.; Shahak, Y.; Aviv-Sharon, E.; Achiam, M.; Gilad, Z. Effects of daytime intra-canopy LED illumination on photosynthesis and productivity of bell pepper grown in protected cultivation. *Sci. Hortic.* **2019**, *250*, 81–88. [CrossRef]
20. Paucek, I.; Pennisi, G.; Pistillo, A.; Appolloni, E.; Crepaldi, A.; Calegari, B.; Spinelli, F.; Cellini, A.; Gabarrell, X.; Orsini, F. Supplementary LED interlighting improves yield and precocity of greenhouse tomatoes in the Mediterranean. *Agronomy* **2020**, *10*, 1002. [CrossRef]
21. Maeda, K.; Masuda, E.; Tamashiro, T.; Maharjan, G.; Maruo, T. Comparison of Supplemental LED Top-and Interlighting for Year-Round Production of Cherry Tomato. *Agronomy* **2022**, *12*, 1878. [CrossRef]
22. Gómez, C.; Clark, M.; Mitchell, C. Effect of intracanopy lighting and/or root-zone temperature on high-wire tomato production under supra-optimal air temperature. *Acta Hortic.* **2016**, *1134*, 63–70. [CrossRef]
23. Verheul, M.; Maessen, H.F.R.; Panosyan, A.; Naseer, M.; Paponov, M.; Paponov, I. Effects of supplemental lighting and temperature on summer production of tomato in Norway. *Acta Hortic.* **2020**, *1296*, 707–714. [CrossRef]
24. Appolloni, E.; Orsini, F.; Pennisi, G.; Gabarrell Durany, X.; Paucek, I.; Gianquinto, G. Supplemental LED lighting effectively enhances the yield and quality of greenhouse truss tomato production: Results of a meta-analysis. *Front. Plant Sci.* **2021**, *12*, 596927. [CrossRef] [PubMed]
25. Kim, Y.-B.; Bae, J.-H.; Park, M.-H. Effects of supplemental lighting on growth and yield of sweet pepper (*Capsicum annuum* L.) in hydroponic culture under low levels of natural light in winter. *Hortic. Sci. Technol.* **2011**, *29*, 317–325.
26. Lee, J.-W.; Kim, H.C.; Jeong, P.H.; Ku, Y.-G.; Bae, J.H. Effects of supplemental lighting of high pressure sodium and lighting emitting plasma on growth and productivity of paprika during low radiation period of winter season. *Hortic. Sci. Technol.* **2014**, *32*, 346–352. [CrossRef]
27. Yoon, S.; Kim, J.H.; Hwang, I.; Kim, D.; Shin, J.; Son, J.E. Effect of stem number on growth, fruit quality, and yield of sweet peppers grown in greenhouses under supplemental lighting with high pressure sodium lamps in winter. *J. Bio-Environ. Control* **2021**, *30*, 237–243. [CrossRef]
28. Kusuma, P.; Pattison, P.M.; Bugbee, B. From physics to fixtures to food: Current and potential LED efficacy. *Hortic. Res.* **2020**, *7*, 56. [CrossRef]
29. Zhang, Z.; Shen, L.; Zhang, H.; Ding, L.; Shao, G.; Liang, X.; Xiang, W. Novel red-emitting CsPb1− xTixI3 perovskite QDs@ glasses with ambient stability for high efficiency white LEDs and plant growth LEDs. *Chem. Eng. J.* **2019**, *378*, 122125. [CrossRef]
30. Lee, J.-G.; Lee, G.J.; Hong, S.C.; Ko, J.-H.; Park, T.; Ko, Y.W. Shape optimization of quantum-dot caps for high color-rendering white light-emitting diodes studied by optical simulation. *J. Korean Phys. Soc.* **2021**, *78*, 822–828. [CrossRef]
31. Jokinen, K.; Särkkä, L.; Näkkilä, J. Improving sweet pepper productivity by LED interlighting. *Acta Hortic.* **2012**, *956*, 59–66. [CrossRef]
32. Brehm, G.; Niermann, J.; Jaimes Nino, L.M.; Enseling, D.; Jüstel, T.; Axmacher, J.C.; Warrant, E.; Fiedler, K. Moths are strongly attracted to ultraviolet and blue radiation. *Insect Conserv. Divers.* **2021**, *14*, 188–198. [CrossRef]
33. Choi, I.-L.; Lee, Y.B.; Kim, I.S.; Kang, H.-M. A comparison of the storability in MA storage and the quality of paprika fruit among cultivars. *J. Bio-Environ. Control* **2012**, *21*, 252–260.

34. Wada, T.; Ikeda, H.; Matsushita, K.; Kambara, A.; Hirai, H.; Abe, K. Effects of Shading in Summer on Yield and Quality of Tomatoes Grown on a Single-truss System. *J. Jpn. Soc. Hortic. Sci.* **2006**, *75*, 51–58. [CrossRef]
35. Lee, J.-N.; Lee, E.-H.; Im, J.-S.; Kwon, Y.-S.; Jang, S.-W.; Yong, Y.-R. Productivity and Fruit Quality according to Training Methods and Harvesting Bate on Paprika during Summer Culture in Highland. *J. Bio-Environ. Control* **2008**, *17*, 204–209.
36. Tilahun, S.; Paramaguru, P.; Rajamani, K. Capsaicin and ascorbic acid variability in chilli and paprika cultivars as revealed by HPLC analysis. *J. Plant Breed. Genet.* **2013**, *1*, 85–89.
37. Kim, D.; Son, J.E. Adding far-red to red, blue supplemental light-emitting diode interlighting improved sweet pepper yield but attenuated carotenoid content. *Front. Plant Sci.* **2022**, *13*, 2041. [CrossRef]
38. Gautier, H.; Diakou-Verdin, V.; Bénard, C.; Reich, M.; Buret, M.; Bourgaud, F.; Poëssel, J.L.; Caris-Veyrat, C.; Génard, M. How does tomato quality (sugar, acid, and nutritional quality) vary with ripening stage, temperature, and irradiance? *J. Agric. Food Chem.* **2008**, *56*, 1241–1250. [CrossRef]
39. Zushi, K.; Suehara, C.; Shirai, M. Effect of light intensity and wavelengths on ascorbic acid content and the antioxidant system in tomato fruit grown in vitro. *Sci. Hortic.* **2020**, *274*, 109673. [CrossRef]
40. Appolloni, E.; Paucek, I.; Pennisi, G.; Stringari, G.; Gabarrell Durany, X.; Orsini, F.; Gianquinto, G. Supplemental LED Lighting Improves Fruit Growth and Yield of Tomato Grown under the Sub-Optimal Lighting Condition of a Building Integrated Rooftop Greenhouse (i-RTG). *Horticulturae* **2022**, *8*, 771. [CrossRef]
41. Kim, H.-J.; Yang, T.; Choi, S.; Wang, Y.-J.; Lin, M.-Y.; Liceaga, A.M. Supplemental intracanopy far-red radiation to red LED light improves fruit quality attributes of greenhouse tomatoes. *Sci. Hortic.* **2020**, *261*, 108985. [CrossRef]
42. Palmitessa, O.D.; Leoni, B.; Montesano, F.F.; Serio, F.; Signore, A.; Santamaria, P. Supplementary far-red light did not affect tomato plant growth or yield under mediterranean greenhouse conditions. *Agronomy* **2020**, *10*, 1849. [CrossRef]
43. Chen, S.; Marcelis, L.F.; Heuvelink, E. Far-red radiation increases flower and fruit abortion in sweet pepper (*Capsicum annuum* L.). *Sci. Hortic.* **2022**, *305*, 111386. [CrossRef]
44. Kotilainen, T.; Aphalo, P.J.; Brelsford, C.C.; Böök, H.; Devraj, S.; Heikkilä, A.; Hernández, R.; Kylling, A.; Lindfors, A.; Robson, T.M. Patterns in the spectral composition of sunlight and biologically meaningful spectral photon ratios as affected by atmospheric factors. *Agric. For. Meteorol.* **2020**, *291*, 108041. [CrossRef]
45. Terashima, I.; Fujita, T.; Inoue, T.; Chow, W.S.; Oguchi, R. Green light drives leaf photosynthesis more efficiently than red light in strong white light: Revisiting the enigmatic question of why leaves are green. *Plant Cell Physiol.* **2009**, *50*, 684–697. [CrossRef]
46. Jang, D.-C.; Choi, K.-Y.; Heo, J.-Y.; Kim, I.-S. Comparison of growth and fruit setting characteristics for selecting the optimum winter-planted paprika cultivars. *Hortic. Sci. Technol.* **2016**, *34*, 424–432. [CrossRef]
47. Yeo, K.-H.; Park, S.H.; Yu, I.-H.; Lee, H.J.; Wi, S.H.; Cho, M.C.; Lee, W.M.; Huh, Y.C. Cultivation demonstration of paprika (*Capsicum annuum* L.) cultivars using the large single-span plastic greenhouse to overcome high temperature in South Korea. *J. Bio-Environ. Control* **2021**, *30*, 429–440. [CrossRef]
48. Britton, C.; Dodd, J. Relationships of photosynthetically active radiation and shortwave irradiance. *Agric. Meteorol.* **1976**, *17*, 1–7. [CrossRef]
49. Jacovides, C.; Tymvios, F.; Asimakopoulos, D.; Theofilou, K.; Pashiardes, S. Global photosynthetically active radiation and its relationship with global solar radiation in the Eastern Mediterranean basin. *Theor. Appl. Climatol.* **2003**, *74*, 227–233. [CrossRef]
50. Lee, H.-W.; Lee, S.; Lee, S. Relationship between total solar radiation and PPF and transmittance in greenhouse at different weather conditions. *J. Bio-Environ. Control* **2002**, *11*, 56–60.
51. Dorais, M. The use of supplemental lighting for vegetable crop production: Light intensity, crop response, nutrition, crop management, cultural practices. In Proceedings of the Canadian Greenhouse Conference, Niagara Falls, ON, Canada, 9 October 2003.
52. Lee, J.; Moon, T.; Du Sung Nam, K.S.P.; Son, J.E. Estimation of leaf area in Paprika based on leaf length, leaf width, and node number using regression models and an artificial neural network. *Hortic. Sci. Technol.* **2018**, *36*, 183–192. [CrossRef]
53. Jang, D.-C.; Choi, K.-Y.; Heo, J.-Y.; Kim, I.-S. The effect of transplant age on growth and fruit yield in winter-planted paprika cultivation. *Hortic. Sci. Technol.* **2018**, *36*, 470–477. [CrossRef]
54. Chen, J.-J.; Zhen, S.; Sun, Y. Estimating leaf chlorophyll content of buffaloberry using normalized difference vegetation index sensors. *HortTechnology* **2021**, *31*, 297–303. [CrossRef]
55. Yoon, H.S.; Choi, I.-L.; Heo, J.-Y.; Kim, J.Y.; Han, S.J.; Kang, H.-M. Influence of hot water immersion and MAP pre-treatments on sterilization and asparagus spear qualities during cold storage. *Hortic. Sci. Technol.* **2018**, *36*, 756–765. [CrossRef]
56. Cunniff, P.; Washington, D. Official methods of analysis of aoac international. *J. AOAC Int.* **1997**, *80*, 127A.
57. Ribes-Moya, A.M.; Raigón, M.D.; Moreno-Peris, E.; Fita, A.; Rodríguez-Burruezo, A. Response to organic cultivation of heirloom Capsicum peppers: Variation in the level of bioactive compounds and effect of ripening. *PLoS ONE* **2018**, *13*, e0207888. [CrossRef]
58. Hwang, H.S.; Lee, K.H.; Jeong, H.W.; Hwang, S.J. Selection of Supplemental Light Source for Greenhouse Cultivation of Pepper during Low Radiation Period through Growth and Economic Analysis. *J. Bio-Environ. Control* **2022**, *31*, 204–211. [CrossRef]

plants MDPI

Article

Shift in the Light Quality of Night Interruption Affects Flowering and Morphogenesis of *Petunia hybrida*

Yoo Gyeong Park [1] and Byoung Ryong Jeong [1,2,3,*]

[1] Institute of Agriculture and Life Science, Gyeongsang National University, Jinju 52828, Republic of Korea; ygpark615@gmail.com
[2] Division of Applied Life Science (BK21 Four), Graduate School, Gyeongsang National University, Jinju 52828, Republic of Korea
[3] Research Institute of Life Science, Gyeongsang National University, Jinju 52828, Republic of Korea
* Correspondence: brjeong@gmail.com; Tel.: +82-010-6751-5489

Abstract: *Petunia hybrida* Hort. "Easy Wave Pink", a qualitative long-day plant (LDP), was investigated to study the effects of the night interruption light (NIL) provided by light-emitting diodes (LEDs) quality shifting on the morphogenesis, blooming, and transcription of photoreceptor genes. Plants were grown in a closed-type plant factory employing white (W) LEDs at an intensity of 180 $\mu mol \cdot m^{-2} \cdot s^{-1}$ PPFD provided for short day (SD, 10 h light, 14 h dark), long day (LD, 16 h light, 8 h dark), or SD with 4 h night interruption (NI) with LEDs at an intensity of 10 $\mu mol \cdot m^{-2} \cdot s^{-1}$ PPFD. The NIL quality was shifted from one light spectrum to another after the first 2 h of NI. Light treatments consisting of all possible pairings of W, far-red (Fr), red (R), and blue (B) light were tested. The SD and LD were referenced as the control, while 12 NI treatments involved altering LED NIL qualities, as follows: from R to B (NI-RB), from B to R (NI-BR), from Fr to R (NI-FrR), from R to Fr (NI-RFr), from Fr to B (NI-FrB), from B to Fr (NI-BFr), from B to W (NI-BW), from W to B (NI-WB), from W to Fr (NI-WFr), from Fr to W (NI-FrW), from W to R (NI-WR), and from R to W (NI-RW). The NI-RFr resulted in the longest shoots, while the NI-WR and NI-RW resulted in the shortest shoots. NI-WR, NI-RW, NI-BW, NI-WB, NI-RFr, NI-RB, NI-BR, and LD all exhibited flowering. High-level expressions of photoreceptor genes were confirmed in the NI-RFr, NI-FrR, NI-BFr, NI-RW, and NI-WR treatments. Morphogenesis and blooming were both impacted by the photoperiod. The first NIL had no effects on the flowering or the morphogenesis, but the second NIL had a profound impact on both.

Keywords: blooming; cryptochrome; photomorphogenesis; phytochrome; wavelength

Citation: Park, Y.G.; Jeong, B.R. Shift in the Light Quality of Night Interruption Affects Flowering and Morphogenesis of *Petunia hybrida*. *Plants* **2023**, *12*, 2049. https://doi.org/10.3390/plants12102049

Academic Editors: Hail Rihan, Mick Fuller and Gianluca Caruso

Received: 12 February 2023
Revised: 26 April 2023
Accepted: 19 May 2023
Published: 21 May 2023

1. Introduction

The light environment (photoperiod, light quality, light intensity, etc.) significantly affects a plant's photomorphogenic development, including flowering, seed germination, and shoot architecture [1,2]. In photoperiodic plants, the phytochrome photoreceptors regulate perception of the light quality, stem extension, and flowering [3]. Plants identify the light quality via photoreceptors, which are categorized as phytochromes, cryptochromes, phototropins, members of the Zeitlupe (ZTL/FKF1/LKP2) family, and the ultraviolet (UV) light receptor(s) [4,5]. Phytochromes are photoreceptors that primarily absorb red (R) and far-red (Fr) light, while cryptochromes absorb UV-A and blue (B) light, both of which help plants bloom [6]. Phototropins play important roles in phototropism, changes in chlorophyll light-avoidance and accumulation movements, inhibition of rapid elongation of the hypocotyl growth, stomatal opening and closing, and leaf expansion [7].

Early flowering can be induced for commercial horticulture businesses by manipulating the photoperiodic conditions [8]. During short-day (SD) seasons, night interruption (NI) is successful in delaying short-day plants (SDPs) from flowering in similarly as to naturally long-day (LD) conditions do, and in hastening long-day plants (LDPs) to flower to allow for earlier sale or seed production [9]. A NI with a very low (3–5 $\mu mol\ m^{-2}\ s^{-1}$ PPFD) light intensity promotes flowering induction and increases growth rates during the juvenile stage in *Cymbidium aloifolium* [10]. In *Arabidopsis thaliana*, an LDP, it was discovered that R light NI was the most effective at preventing flower blooming, and that this inhibition was frequently reversible with Fr light [11]. In *Petunia hybrida*, an LDP, NI treatments with green, Fr, and white (W) light encouraged flowering [12]. According to Shin et al. [13], NI treatment with a combination of B and R light encouraged *Cyclamen persicum* flowering in the winter. In *Pelargonium* × *hortorum*, a day-neutral plant (DNP), NI treatment with B, green, R, Fr, and W light encouraged flowering but delayed it in NI treatment with Fr light [14]. By keeping herbaceous SDPS in their vegetative growth stage, NI was also employed to prevent or delay flowering in *Dendranthema grandiflorum* [15,16] and *Kalanchoe blossfeldiana* [11]. *K. blossfeldiana* 'Lipstick' (SDP) was not affected in flowering by any of the night interruption lights (NILs), such as B, R, W, or combination of B and W, while the 'Spain' variety flowered only in the 10 μmol m^{-2} s^{-1} PPFD NI-interruptional B light [17]. *D. grandiflorum* (SDP) responded the most strongly to R light NI for inhibiting flowering, but NI with Fr light or B light had less of an impact [18]. According to Yang et al. [19], 30 μmol m^{-2} s^{-1} PPFD supplemental B light and NI-interruptional B light were more effective in promoting growth, flowering, and the expression of florigen genes. Plant responses to light quality are species-specific, according to research conducted under various light environments [20].

However, NI using LEDs with varying light qualities was only tested on *D. grandiflorum* (SDP) [21] and *Pelargonium* × *hortorum* (DNP) [22] and not on LDPs. Our previous study on shifts in light quality of NI in SDP [21] and DNP [22] showed that morphogenesis and flowering were affected by the second NIL, but the first NIL had no effects on either. Flowering of SDP was observed in the NI-RB, NI-FrR, NI-BFr, NI-FrB, NI-WB, NI-FrW, NI-WFr, NI-WR, and SD treatments, and was especially promoted in the NI-BFr and NI-FrB treatments [21]. DNP was observed to flower in all NI treatments, and flowering was promoted in NI-RFr and NI-FrR treatments [22]. We have reported the shifting of NIL in SDP and DNP; our findings on LDP are also important. We hypothesized that the first and second NIL would affect morphogenesis and flowering in LDP. Furthermore, new practical applications of NIL quality shifting for the floricultural industry would be of substantial interest. The effects on *P. hybrida* Hort. "Easy Wave Pink", an LDP, of NIL quality shifting on the blooming, morphogenesis, and transcription of the photoreceptor genes was therefore investigated in this work.

2. Results

2.1. Morphogenesis

NI-Fr resulted in the longest shoots, while NI-WR and NI-RW resulted in the shortest shoots (Figure 1). In comparison to that in LD, the leaf-length-to-leaf-width ratio dropped in the NI-WB, NI-BW, and NI-RW treatments (Figure 2A). The ratio of leaf length to petiole length grew in the NI-WB, NI-RB, NI-BW, and NI-WFr treatments as compared to the LD treatment (Figure 2B). In contrast, Fr light-driven decrease in the leaf-length-to-petiole-length ratio as seen in NI-FrW, NI-FrB, NI-BFr, NI-FrR, and NI-RFr, suggests succulent growth. NI-BR resulted in the greatest number of leaves per plant, followed by NI-WB and NI-RW (Figure 2C). The leaf area improved in all NI treatments and SD compared to that in LD, probably due to inhibition of flowering in petunia and continued vegetative growth in those deceptive photoperiods (Figure 2D).

The relative growth rate was the greatest in the NI-WB and the least in the SD treatments (Figure 3). With the exception of NI-BFr, all NI treatments led to higher chlorophyll content. The greatest chlorophyll content was found in NI-RB, NI-BR, NI-WB, and SD (Figure 4). The shoot fresh weight was the greatest in NI-BR, the shoot dry weight was the greatest in Ni-WB, and considerably higher biomass was obtained in all NI treatments (Table 1).

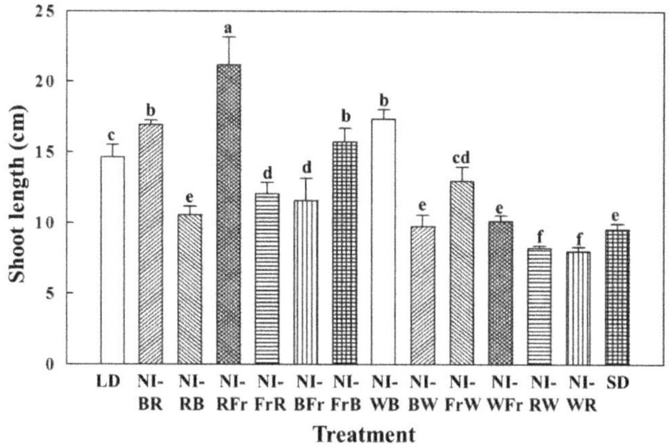

Figure 1. The effects of the NIL quality shifting at 10 $\mu mol \cdot m^{-2} \cdot s^{-1}$ PPFD on the shoot length of petunia (*Petunia hybrida* Hort. 'Easy Wave Pink'), taken 66 days after treatment: NI-BR, blue to red; NI-RB, red to blue; NI-RFr, red to far-red; NI-FrR, far-red to red; NI-BFr, blue to far-red; NI-FrB, far-red to blue; NI-WB, white to blue; NI-BW, blue to white; NI-FrW, far-red to white; NI-WFr, white to far-red; NI-RW, red to white; and NI-WR, white to red. The LD indicates the 16 h long-day treatment. Vertical bars indicate means ± S.E. (n = 3). Means accompanied by different letters are significantly different ($p < 0.05$) according to the Duncan's multiple range test at 5% significance level.

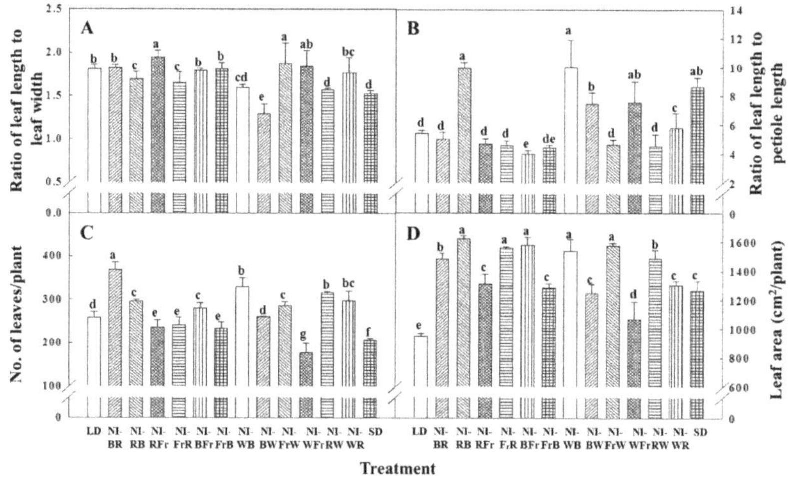

Figure 2. The effects of the NIL quality shifting at 10 $\mu mol \cdot m^{-2} \cdot s^{-1}$ PPFD on the leaf-length-to-leaf-width ratio (**A**), leaf length to petiole length ratio (**B**), average number of leaves (**C**), and leaf area (**D**) of petunia (*Petunia hybrida* Hort. 'Easy Wave Pink') taken 66 days after treatment. Please refer to Figure 1 for detailed NIL qualities. Vertical bars indicate means ± S.E. (n = 3). Means accompanied by different letters are significantly different ($p < 0.05$) according to the Duncan's multiple range test at 5% significance level.

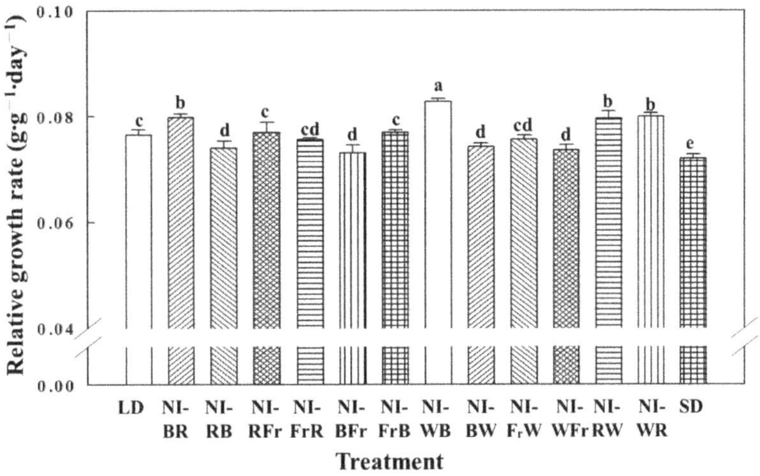

Figure 3. The effects of the NIL quality shifting at 10 μmol·m^{-2}·s^{-1} PPFD on the relative growth rate of petunia (*Petunia hybrida* Hort. 'Easy Wave Pink') taken 66 days after treatment. Please refer to Figure 1 for detailed NIL qualities. Vertical bars indicate means ± S.E. (n = 3). Means accompanied by different letters are significantly different ($p < 0.05$) according to Duncan's multiple range test at 5% significance level.

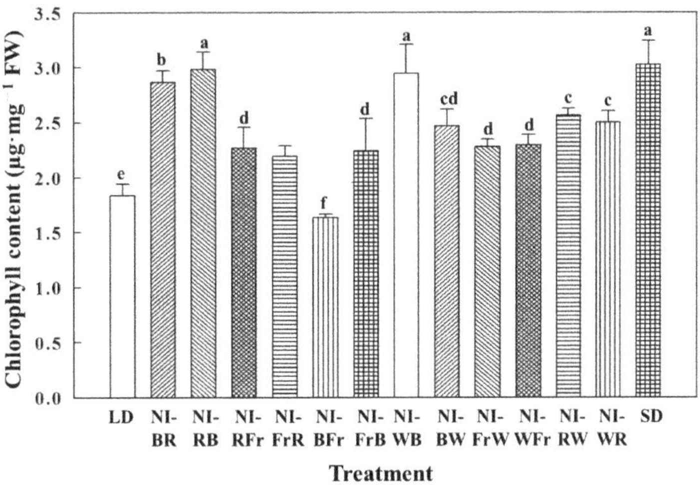

Figure 4. The effects of the NIL quality shifting at 10 μmol·m^{-2}·s^{-1} PPFD on the chlorophyll content of petunia (*Petunia hybrida* Hort. 'Easy Wave Pink') leaves, taken 66 days after treatment. Please refer to Figure 1 for detailed NIL qualities. Vertical bars indicate means ± S.E. (n = 3). Means accompanied by different letters are significantly different ($p < 0.05$) according to Duncan's multiple range test at 5% significance level.

Table 1. The effects of the NIL quality shifting at 10 μmol·m^{-2}·s^{-1} PPFD on the shoot and root fresh and dry weights in petunia (*Petunia hybrida* Hort. 'Easy Wave Pink'), taken 66 days after treatment.

Treatment [z]	Fresh Weight (g)			Dry Weight (g)		
	Shoot	Root	Total	Shoot	Root	Total
LD	64.8 bc [y]	3.77 de	68.6 b–d	5.04 cd	0.35 e	5.37 cd
NI-BR	75.2 a	3.47 e	78.7 a	5.97 b	0.32 e	6.29 b
NI-RB	68.0 ab	4.96 c	73.0 a–c	4.45 d–f	0.36 de	4.82 c–e
NI-RFr	57.5 cd	4.35 c–e	61.9 de	5.08 cd	0.45 c	5.54 bc
NI-FrR	61.2 bc	5.82 b	67.0 b–d	4.57 d–f	0.59 b	5.16 c–e
NI-BFr	58.4 cd	6.11 ab	64.6 c–e	4.16 ef	0.46 c	4.63 de
NI-FrB	57.7 cd	6.67 a	64.4 c–e	4.85 de	0.67 a	5.52 bc
NI-WB	68.9 ab	6.73 a	75.6 ab	6.75 a	0.48 c	7.24 a
NI-BW	57.4 cd	4.68 c	62.1 de	4.51 d–f	0.34 e	4.85 c–e
NI-FrW	63.2 bc	3.67 de	66.8 b–d	4.85 de	0.32 e	5.17 c–e
NI-WFr	52.0 d	4.06 c–e	56.1 e	4.26 d–f	0.44 cd	4.70 c–e
NI-RW	64.7 bc	4.33 c–e	69.0 b–d	5.82 bc	0.43 cd	6.25 b
NI-WR	64.4 bc	4.53 cd	69.0 b–d	5.90 b	0.44 cd	6.34 b
SD	51.5 d	4.88 c	56.4 e	3.88 f	0.49 c	4.37 e
F-test	***	***	***	***	***	***

[z] Please refer to Figure 1 for the detailed NIL qualities. [y] Mean separation within columns by Duncan's multiple range test at a 5% level. ***: Significant at $p \leq 0.001$, respectively.

2.2. *Flowering*

The percentage of flowered buds observed was as follows: 100% in the NI-RW, NI-BR, and LD treatments; 75% in the NI-RFr and NI-WB treatments; and 50% in the NI-WR, NI-BW, and NI-RB treatments (Table 2 and Figure 5). The DVB increased in those NI treatments in which the plant flowered as compared to the LD (Table 2). Flowering results observed in this experiment are also a reflection of the R-to-Fr light ratio, and the effects of quality shifting on flowering was more pronounced by the second NIL than the first NIL.

Table 2. The effects of the NIL quality shifting at 10 μmol·m^{-2}·s^{-1} PPFD on the flowering characteristics of petunia (*Petunia hybrida* Hort. 'Easy Wave Pink'), taken 66 days after treatment.

Treatment [z]	Flowering (%)	DVB [y] (Day)	No. of Flowers/Plant
LD	100	31.0	49.6 a [x]
NI-BR	100	62.6	20.3 b
NI-RB	50	35.0	4.0 cd
NI-RFr	75	59.6	4.3 cd
NI-FrR	- [w]	-	-
NI-BFr	-	-	-
NI-FrB	-	-	-
NI-WB	75	73.6	7.3 c
NI-BW	50	50.6	4.3 cd
NI-FrW	-	-	-
NI-WFr	-	-	-
NI-RW	100	71.3	6.3 cd
NI-WR	50	50.6	3.6 cd
SD	-	-	-
F-test			NS [v]

[z] Please refer to Figure 1 for the detailed NIL qualities. [y] Days after treatment initiation to visible flower bud or days to visible buds. [x] Mean separation within columns by Duncan's multiple range test at a 5% level. [w] No flowering. [v] NS: Nonsignificant.

High-level expressions of photoreceptor genes were confirmed in the NI-RFr, NI-FrR, NI-BFr, NI-RW, and NI-WR treatments (Figure 6). Compared to those in LD, phytochromes (*phyA* and *phyB*) were extensively expressed at elevated levels in the NI-RFr and NI-FrR treatments (Figure 6). *FTL* genes were studied under different light conditions during the

NI along with the LD and SD, and compared to other conditions, the expression of *FTL* and *AFT* were more substantial in the NI-RFr and NI-FrR treatments.

Figure 5. The effects of the NIL quality shifting at 10 μmol·m^{-2}·s^{-1} PPFD on flowering of petunia (*Petunia hybrida* Hort. 'Easy Wave Pink'), taken 66 days after treatment side view (**A**) and top view (**B**). Please refer to Figure 1 for the detailed NIL qualities.

Figure 6. The effects of the NIL quality shifting at 10 μmol·m^{-2}·s^{-1} PPFD on the expression of photoreceptor genes in petunia (*Petunia hybrida* Hort. 'Easy Wave Pink'), taken 66 days after treatment. Please refer to Figure 1 for the detailed NIL qualities. The *phytochrome A* (*phyA*), *phytochrome B* (*phyB*), *cryptochrome 1* (*cry 1*), *Anti-florigenic FT/TFL1* family protein (*AFT*), and *FLOWERING LOCUS T* (*FTL*) are all the indicated photoreceptor genes. Comparative gene expressions were conducted using a constitutively expressed *Actin* gene as the control.

3. Discussion

Numerous elements of plant development, including flowering, stem elongation, and seed germination are controlled by the plant hormone gibberellic acid [23] and light [24]. Increased shoot length observed in this study indicates the involvement of gibberellic acid as affected by the Fr light used in the NI. This study's results show that NI-RFr caused the formation of the longest shoots, while NI-WR and NI-RW formed the shortest shoots, indicating that Fr light encouraged shoot extension while R and W lights inhibited it.

Following exposure to Fr-rich light, many plant species greatly speed their elongation within minutes [23]. On the other hand, exposing plants to R-rich light again causes the extension to slow down to the same degree. Phytochromes are involved in such R/Fr light reversibility. Fr-rich light can accelerate flowering, reduce assimilate accumulation, reduce seed set, decrease fruit development, and impair seed quality, while increasing the elongation, which coincides with stronger apical dominance and reduced branching [25]. The increased shoot length observed in the NI-RFr treatment in this study might have been related to the effects of Fr light given in the second or last NIL. Furthermore, suppressed shoot length in both the NI-RW and NI-WR treatments might have been due to the relatively greater leaf expansion observed in the treatments with NIL composed of R and W lights.

The leaf-length-to-leaf-width-ratio decreased further in the NI-RW, NI-WB, and NI-BW treatments compared to that in the LD treatment, indicating relatively greater leaf expansion induced by the W light. The leaf area increased in all NI treatments and SD as compared to the LD treatment, probably due to inhibition of flowering in petunia and continued vegetative growth in those unreceptive photoperiods.

The greatest chlorophyll content was observed in SD, NI-RB, NI-BR, and NI-WB, which was probably enhanced by B light, especially given that, in the last NI period, the B light quickly and reversibly regulates the stomatal aperture, which enables a great stomatal aperture [26]. It is quite plausible that zeaxanthin [27], cryptochromes, and phototropins [28] are also involved in the B light signaling in the guard cells. Increased chlorophyll content in those treatments may have caused more active photosynthesis, resulting in enhanced biomass prompted by the NI even at low light intensities.

In this study, the percentages of flowered buds observed were as follows: 100% in the NI-WR, NI-BR, and LD treatments; 75% in the NI-RFr and NI-WB treatments; and 50% in the NI-RW, NI-BW, and NI-RB treatments. Our previous study [12] suggested that percentage of flowering was greatest in LD (100%), followed by both NI-R (33.3%) and NI-Fr (33.3%) and both NI g (16.6%) and NI-W (16.6%) during SD with 4 h NI treatment. In our previous study, R, Fr, G, and W light induced flowering in petunia [12], but in this study with shift in 2 h NIL each, B light also affected flowering induction. This suggests that the second NIL quality has a greater effect on flowering induction than the first NIL quality due to the complexity of shift in NIL quality.

The phytochrome photoequilibrium, which affects flowering of photoperiodic crops, is influenced by the R-to-Fr light ratio (P_{Fr}/P_{R+Fr}) [3]. Flowering and stem extension in LDPs is promoted by a low P_{Fr}/P_{R+Fr} [29]. Artificial lighting with Fr light, especially at the conclusion of the photoperiod, drives many LDPs to flower the most quickly [30,31]. *A. thaliana* flowered later when exposed continuously to high P_{Fr}/P_{R+Fr} light than when grown with a low P_{Fr}/P_{R+Fr} light [32]. Additionally, plants flowered much later when grown with continuous R light than when grown with continuous B light [33]. Flowering results of the petunia observed in this study are also the reflection of the P_{Fr}/P_{R+Fr}, and the effects of quality shifting on flowering was more pronounced by the second than the first NIL. In the model plant *A. thaliana*, the roles of *phyA* and *phyB* have been thoroughly explored [4]. It has been observed that plants under continuous Fr and R light with the *phyA* and *phyB* mutants grow taller than the wild type [34], and shows how *phyA* and *phyB* function to detect the appropriate wavelength of light to trigger the hypocotyl inhibitory response. In this study, the phytochromes (*phyA* and *phyB*) in petunia were extensively expressed at higher levels in the NI-RFr compared to in the LD. This implies that phytochromes (*phyA* and *phyB*) were involved in the Fr and R light perception for initiation of flowering in the petunia as also described previously in *A. thaliana* [35]. The roles of cryptochromes in inducing flowering have been observed by utilizing mutations in the *cry1* and *cry2* genes, on the other hand, because cryptochromes are known to stimulate flowering. In this study, the effects of the NIL quality, along with the LD and SD, were investigated, because it is well known that the LD and SD extensively affect flowering in *A. thaliana*. In the NI-RFr, NI-WR, and NI-RW treatments, since a high level of *cry1* gene expression was induced, the plants flowered, but not in the NI-FrR treatment. Flowering plants induced the expression of the *cry1* gene in NI-RFr, which explains that cryptochromes were receptors of these light conditions. In the NI-FrR treatment, even though a high level of *phyA*, *cry1*, and *FTL* gene expression was induced, the plants did not flower. In the NI-BFr treatment, even though *phyA* and *FTL* were highly expressed, the plants did not flower. Indeed, phenotypic studies also showed that the petunia had the receptors of the B, W, Fr, and R light even under photoperiods other than LD by showing blooming of flowers.

In addition, the genes similar to *FTL* and *TFL* play an important part in integrating endogenous and exogenous flowering-controlling signals [36]. The *FTL* and *TFL* encode small globular-like proteins. In this study, *FTL* genes were studied under different light conditions during the NI along with the LD and SD, and the expression of these genes (*FTL* and *AFT*) were more pronounced in NI-RFr, NI-FrR, and NI-BFr than in other conditions. This indicates that the *FTL* was also the receptor of the B, R, and Fr, in addition to the LD and SD. Overall, this study presumed that the B, Fr, and R light was mainly received by phytochromes, cryptochromes, and flowering terminal locus genes in the petunia.

4. Materials and Methods

4.1. Plant Materials and Growth Conditions

Petunia (Pan Seed Co., West Chicago, IL, USA) seedlings were transplanted from a glasshouse bench into 50-cell plug trays with a commercial medium (Tosilee Medium, Shinan Grow Co., Jinju, Republic of Korea) 40 days after sowing. On the day of transplanting, the seedlings and rooted cuttings were moved to a closed-type plant factory. After settling in for 24 days in the plant factory, the plants with shoot lengths of around 6.0 cm were exposed to the photoperiodic light treatments. For LDPs, the critical day length is 14 h, and for SDPs, it is 12 h. The plants were grown in a glass house and then moved to the plant factory, first to adapt to 20 ± 1 °C, 60 ± 10% RH, and 140 ± 20 $\mu mol \cdot m^{-2} \cdot s^{-1}$ PPFD from fluorescent lamps (F48T12-CW-VHO, Philips Co., Ltd., Eindhoven, The Netherlands) and subsequently acclimatize to the photoperiodic treatments provided by LED lighting systems at 25 cm atop the plant canopy. Throughout the experiment, the petunia was fertigated once every day with a greenhouse multipurpose nutrient solution [in $mg \cdot L^{-1}$ $Ca(NO_3)_2 \cdot 4H_2O$ 737.0, KNO_3 343.4, KH_2PO_4 163.2, K_2SO_4 43.5, $MgSO_4 \cdot H_2O$ 246.0, NH_4NO_3 80.0, Fe-EDTA 15.0, H_3BO_3 1.40, $NaMoO_4 \cdot 2H_2O$ 0.12, $MnSO_4 \cdot 4H_2O$ 2.10, and $ZnSO_4 \cdot 7H_2O$ 0.44].

4.2. Photoperiodic Light Treatments

The petunia was cultivated using white LEDs at an intensity of 180 $\mu mol \cdot m^{-2} \cdot s^{-1}$ PPFD for long day (16 h light, 8 h dark), short day (10 h light, 14 h dark), or SD with a 4 h (23:00–3:00) night interruption with LEDs at an intensity of 10 $\mu mol \cdot m^{-2} \cdot s^{-1}$ PPFD. The NIL quality was shifted from one to another after the first 2 h of NI until the end of the experiments for 66 days. The employed NIL qualities in this study were W (400–700 nm), R (660 nm), Fr (730 nm), and B (450 nm) (Figure 7). The SD and LD were referenced as the control in this study, and 12 NI treatments combined with different NIL combinations were studied, formulated as follows: B to R (NI-BR), R to B (NI-RB), R to Fr (NI-RFr), Fr to R (NI-FrR), B to Fr (NI-BFr), Fr to B (NI-FrB), W to B (NI-WB), B to W (NI-BW), Fr to W (NI-FrW), W to Fr (NI-WFr), R to W (NI-RW), and W to R (NI-WR) (Figure 8). A spectroradiometer (USB 2000 Fiber Optic Spectrometer, Ocean Optics Inc., Dunedin, FL, USA) scanned the spectral distribution of lights in all treatments 25 cm above the bench top in 1 nm intervals. In each light treatment, the average of the maximum absolute irradiance, and the spectral distribution were measured at three different locations within the plant-growing bench.

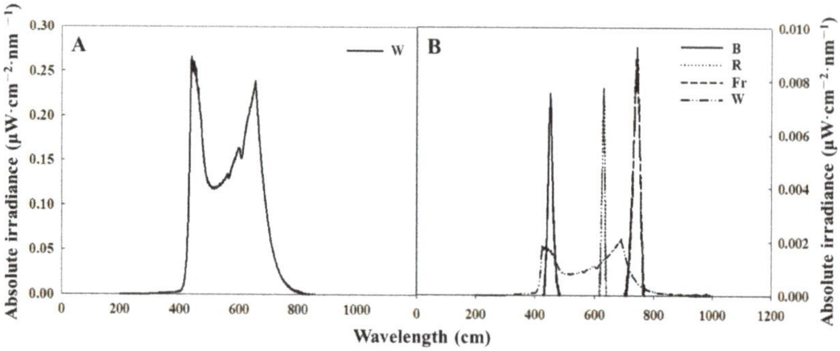

Figure 7. The spectral distribution of lights employed in the closed-type plant factory: White light-emitting diodes (LEDs) used as daily light (**A**) and different LEDs (B, blue; R, red; Fr, far-red; and W, white) used as night interruption light (NIL) provided by LEDs for 2 h each (**B**).

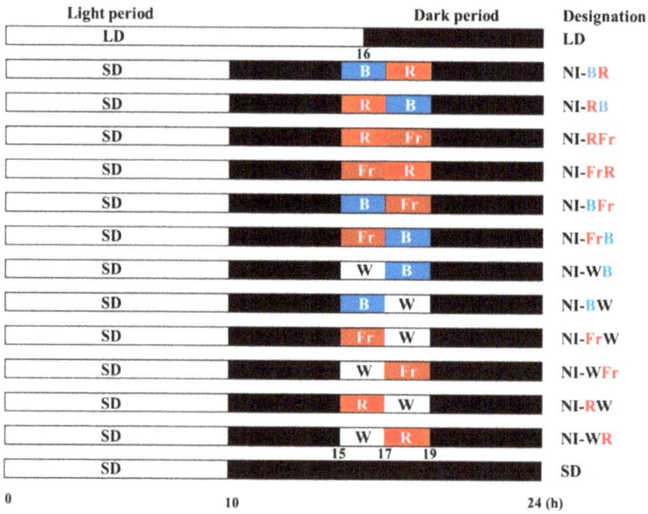

Figure 8. NIL quality shifting using LEDs during the 4 h night interruption (NI) in the 10 h short-day (SD) treatments. Please refer to Figure 1 for detailed NIL qualities.

4.3. Data Collection and Analysis

The leaf length, shoot length, petiole length, leaf width, average number of leaves, chlorophyll content, fresh and dry shoot and root weights, relative growth rate, percent flowering, days from the start of the photoperiodic treatments to the first visible flower bud or days to visible buds (DVB), average number of flowers, and expression of photoreceptor genes were all assessed after 66 days. The leaf expansion index was calculated as the proportion of the leaf length to leaf width, and the overgrowth (stretchiness) index was calculated as the proportion of the leaf length to the petiole length. The mean net increase in the dry biomass divided by the plant dry biomass over a period of time was taken as the relative growth rate. Before (W1) and after (W2) the treatments were applied, the total plant dry weight was measured, and the relative growth rate between finishing (t2) and starting (t1) days of the experiments was calculated as follows:

$$\text{Relative growth rate} = (\ln W2 - \ln W1)/(t2 - t1)$$

In total, 10 mg of fresh, young completely formed leaves were extracted using 80% ice-cold acetone to estimate the chlorophyll content. The absorbance of the supernatant was assessed with a spectrophotometer (Biochrom Libra S22, Biochrom Co., Ltd., Holliston, MA, USA) at 663 and 645 nm after centrifugation at 3000 rpm. Dere et al. [37] was referenced for calculations. After drying for three days at 75 °C in an oven (Model FO-450M, Jeio Technology Co., Ltd., Seoul, Republic of Korea), the shoot and root dry weights were measured.

However, treatment effects was also analyzed separately for the first (first NI) and second (second NI) 2 h periods of the 10 h short-day (SD) treatments based on the assumption that the group of the same light-quality treatments during the same period being the same treatment, e.g., in the first period the blue (B) treatment consisted of B to R (NI-BR), B to Fr (NI-BFr), and B to W (NI-BW).

This experiment used a randomized complete block design, with 3 replications, and 2 plants for each replication. To reduce the effects of the position, the treatment sites in a controlled setting were arbitrarily mixed between replications. The SAS (Statistical Analysis System, V. 9.1, Cary, NC, USA) program was used to determine the statistical significance of the acquired data. Duncan's multiple range test and an analysis of variance (ANOVA) were applied to the results of this experiment. Graphing was completed using Sigma Plot 10.0 (Systat Software, Inc., San Jose, CA, USA).

4.4. Isolation of the Total RNA Isolation and Semi-Quantitative RT-PCR (Reverse Transcriptase-Polymerase Chain Reaction) Analysis of a Subset of Genes

Following the manufacturer's protocols (Promega, Madison, WI, USA), the total RNA was isolated from the shoot tip of the plants that had been exposed to 33 days of NI treatments. Using a reverse transcriptase kit from Promega, Madison, WI, USA, 1 µg of DNase-treated RNA was reverse-transcribed to create first-strand cDNA, which was then utilized as the template for the PCR (polymerase chain reaction). The *phytochrome A* (*phyA*), *phytochrome B* (*phyB*), *cryptochrome 1* (*cry1*), *Anti-florigenic FT/TFL1* family protein (*AFT*), and *FLOWERING LOCUS T* (*FTL*) genes of the sequence from *Arabidopsis thaliana* were used as primers in separate PCRs with an equal amount of cDNA (Table 3). In petunia, similar to Arabidopsis, flowering is delayed under R light and induced under B light; however, its mechanism still remains unknown. Therefore, *A. thaliana* primers with similar gene sequences [38,39] were used. *Actin* was employed as the control because, due to its high conservation as an endogenous housekeeping gene, it is frequently used to normalize molecular expression investigations. The following PCR conditions were used: 5 min initial denaturation for at 95 °C, 35 20 s cycles at 95 °C, 30 s at 57 °C, 30 s at 72 °C, and a final extension step of 10 min at 72 °C. After 35 cycles, the PCR results were tested on a 1% agarose gel to determine whether the transcripts were expressed differently.

Table 3. Primers used to quantify gene expression levels.

Gene	Accession No.	Forward Primer	Reverse Primer
phyA	EU915082	5′-GACAGTGTCAGGCTTCAACAAG-3′	5′-ACCACCAGTGTGTGTTATCCTG-3′
phyB	NM_127435	5′-GTGCTAGGGAGATTACGCTTTC-3′	5′-CCAGCTTCTGAGACTGAACAGA-3′
cry1	NM_116961	5′-CGTAAGGGATCACCGAGTAAAG-3′	5′-CTTTTAGGTGGGAGTTGTGGAG-3′
AFT	AB839766	5′-AGAACACCTCCATTGGATCG-3′	5′-CTGGAACTAGGTGGCCTCAC-3′
FTL	AB839767	5′-ACAACGGACTCCTCATTTGG-3′	5′-CGCGAAACTACGAGTGTTGA-3′
Actin	AB205087	5′-CGTTTGGATCTTGCTGGTCG-3′	5′-CAGGACATCTGAAACGCTCA-3′

5. Conclusions

NI-RFr resulted in the longest shoots, whereas NI-WR and NI-RW resulted in the shortest shoots, which indicates that FR light encouraged shoot extension. LD, NI-BR, NI-RB, NI-WR NI-WB, NI-RFr, NI-RW, and NI-BW caused plants to flower. Photoperiod affected both morphogenesis and flowering. While the first NIL had no effects on flowering or morphogenesis, the second NIL had a profound impact on both.

Author Contributions: Conceptualization, Y.G.P. and B.R.J.; methodology, Y.G.P. and B.R.J.; formal analysis, Y.G.P.; data curation, Y.G.P.; writing—original draft preparation, Y.G.P.; writing—review and editing, Y.G.P. and B.R.J. All authors have read and agreed to the published version of the manuscript.

Funding: This research received no external funding. Yoo Gyeong Park was supported by the BK21 Program, Ministry of Education, Republic of Korea.

Data Availability Statement: Data sharing is not applicable to this article.

Acknowledgments: The authors express their gratitude to Sowbiya Muneer, Prabhakaran Soundararajan, and Young Don Chin for their assistance.

Conflicts of Interest: The authors declare no conflict of interest.

References

1. Cerdán, P.D.; Chory, J. Regulation of flowering time by light quality. *Nature* **2003**, *423*, 881–885. [CrossRef] [PubMed]
2. Takano, M.; Inagaki, N.; Xie, X.; Kiyota, S.; Baba-Kasai, A.; Tanabata, T.; Shinomura, T. Phytochromes are the sole photoreceptors for perceiving red/far-red light in rice. *Proc. Natl. Acad. Sci. USA* **2009**, *106*, 14705–14710. [CrossRef]
3. Craig, D.S.; Runkle, E.S. Using LEDs to quantify the effect of the red to far-red ratio of night-interruption lighting on flowering of photoperiodic crops. *Acta Hortic.* **2012**, *956*, 179–185. [CrossRef]
4. Lin, C.T. Plant blue-light receptors. *Trends Plant Sci.* **2000**, *5*, 337–342. [CrossRef] [PubMed]
5. Takemiya, A.; Inoue, S.; Doi, M.; Kinoshitaa, T.; Shimazaki, K. Phototropins promote plant growth in response to blue light in low light environments. *Plant Cell* **2005**, *17*, 1120–1127. [CrossRef] [PubMed]
6. Cashmore, A.R.; Jarillo, J.A.; Wu, Y.-J.; Liu, D. Cryptochromes: Blue light receptors for plants and animals. *Science* **1999**, *284*, 760–765. [CrossRef] [PubMed]
7. Kimura, M.; Kagawa, T. Phototropin and light-signaling in phototropism. *Curr. Opin. Plant Biol.* **2006**, *9*, 503–508. [CrossRef] [PubMed]
8. Warner, R.M.; Erwin, J.E. Effect of photoperiod and daily light integral on flowering of five *Hibiscus* sp. *Sci. Hortic.* **2003**, *97*, 341–351. [CrossRef]
9. Hamamoto, H.; Hideo, S.; Tadahisa, H. Budding response of horticultural crops to night break with red light on alternate days. *Environ. Control Biol.* **2005**, *43*, 21–27. [CrossRef]
10. Kim, Y.J.; Lee, H.J.; Kim, K.S. Night interruption promotes vegetative growth and flowering of *Cymbidium*. *Sci. Hortic.* **2011**, *130*, 887–893. [CrossRef]
11. Thomas, B.; Vince-Prue, D. *Photoperiodism in Plants*, 2nd ed.; Academic Press: London, UK, 1997.
12. Park, Y.G.; Muneer, S.; Soundararajan, P.; Manivannan, A.; Jeong, B.R. Light quality during night interruption affects morphogenesis and flowering in *Petunia hybrida*, a qualitative long-day plant. *Hortic. Environ. Biotechnol.* **2016**, *57*, 371–377. [CrossRef]
13. Shin, J.H.; Jung, H.H.; Kim, K.S. Night interruption using light emitting diodes (LEDs) promotes flowering of *Cyclamen persicum* in winter cultivation. *Hortic. Environ. Biotechnol.* **2010**, *51*, 391–395.
14. Park, Y.G.; Muneer, S.; Soundararajan, P.; Manivannan, A.; Jeong, B.R. Light quality during night interruption affects morphogenesis and flowering in geranium. *Hort. Environ. Biotechnol.* **2017**, *58*, 212–217. [CrossRef]
15. Ochiai, M.; Liao, Y.; Shimazu, T.; Takai, Y.; Suzuki, K.; Yano, S.; Fukui, H. Varietal differences in flowering and plant growth under night-break treatment with LEDs in 12 chrysanthemum cultivars. *Environ. Control Biol.* **2015**, *53*, 17–22. [CrossRef]
16. Blanchard, M.G.; Runkle, E.S. Use of a cyclic high-pressure sodium lamp to inhibit flowering of chrysanthemum and velvet sage. *Sci. Hortic.* **2009**, *122*, 448–454. [CrossRef]
17. Kang, D.I.; Jeong, H.K.; Park, Y.G.; Jeong, B.R. Flowering and morphogenesis of kalanchoe in response to quality and intensity of night interruption light. *Plants* **2019**, *8*, 90. [CrossRef]
18. Higuchi, Y.; Sumitomo, K.; Oda, A.; Shimizu, H.; Hisamatsu, T. Days light quality affects the night-break response in the short-day plant chrysanthemum, suggesting differential phytochrome-mediated regulation of flowering. *J. Plant Physiol.* **2012**, *169*, 1789–1796. [CrossRef]
19. Yang, J.; Song, J.; Jeong, B.R. The flowering of SDP chrysanthemum in response to intensity of supplemental or night-interruptional blue light is modulated by both photosynthetic carbon assimilation and photoreceptor-mediated regulation. *Front. Plant Sci.* **2022**, *13*, 981143. [CrossRef]
20. Hogewoning, S.; Trouwborst, G.; Maljaars, H.; Poorter, H.; Van Ieperen, W.; Harbinson, J. Blue light dose-responses of leaf photosynthesis, morphology, and chemical composition of *Cucumis sativus* grown under different combinations of red and blue light. *J. Exp. Bot.* **2010**, *61*, 3107–3117. [CrossRef]
21. Park, Y.G.; Muneer, S.; Jeong, B.R. Morphogenesis, flowering, and gene expression of *Dendranthema grandiflorum* in response to shift in light quality of night interruption. *Int. J. Mol. Sci.* **2015**, *16*, 16497–16513. [CrossRef] [PubMed]
22. Park, Y.G.; Jeong, B.R. Photoreceptors modulate the flowering and morphogenesis responses of *Pelargonium x hortorum* to night-interruption light quality shifting. *Agronomy* **2023**, *13*, 857. [CrossRef]
23. Van Ieperen, W. Plant morphological and developmental responses to light quality in a horticultural context. *Acta Hortic.* **2012**, *956*, 131–139. [CrossRef]

24. Halliday, K.J.; Fankhauser, C. Phytochrome-hormonal signaling networks. *New Phytol.* **2003**, *157*, 449–463. [CrossRef] [PubMed]
25. Whitelam, G.C.; Halliday, K.J. Light and plant development. In *Red: Far-Red Ratio Perception and Shade Avoidance*; Franklin, K.A., Whitelam, G.C., Eds.; Blackwell: Oxford, UK, 2007; pp. 211–229.
26. Meidner, H.; Mansfield, T.A. *Physiology of Stomata*, 1st ed.; McGraw-Hill: London, UK, 1968.
27. Zeiger, E.; Talbott, L.D.; Frechilla, S.; Srivastava, A.; Zhu, J. The guard cell chloroplast: A perspective for the twenty-first century. *New Phytol.* **2002**, *153*, 415–424. [CrossRef] [PubMed]
28. Kinoshita, T.; Doi, M.; Suetsugu, N.; Kagawa, T.; Wada, M.; Shimazaki, K. Photo1 and photo 2 mediate blue light regulation of stomatal opening. *Nature* **2001**, *414*, 656–660. [CrossRef] [PubMed]
29. Runkle, E.S.; Heins, R.D. Manipulating the light environment to control flowering and morphogenesis of herbaceous plants. *Acta Hortic.* **2006**, *711*, 51–59. [CrossRef]
30. Downs, R.J.; Thomas, J.F. Phytochrome regulation of flowering in the long-day plant *Hyoscyamus niger*. *Plant Physiol.* **1982**, *70*, 898–900. [CrossRef] [PubMed]
31. Lane, H.C.; Cathey, H.M.; Evans, L.T. Dependence of flowering in several long-day plants on spectral composition of light extending photoperiod. *Am. J. Bot.* **1965**, *52*, 1006–1014. [CrossRef]
32. Halliday, K.J.; Koornneef, M.; Whitelam, G.C. Phytochrome B and at least one other phytochrome mediate the accelerated flowering response of *Arabidopsis thaliana* L. to low red/far-red ratio. *Plant Physiol.* **1994**, *104*, 1311–1315. [CrossRef]
33. Guo, H.; Yang, H.; Mockler, T.C.; Lin, C. Regulation of flowering time by *Arabidopsis* photoreceptors. *Science* **1998**, *279*, 1360–1363. [CrossRef]
34. Koornneef, M.; Rolff, E.; Spruitt, C.J.P. Genetic control of light-inhibited hypocotyl elongation in *Arabidopsis thaliana* L. *Int. J. Plant Physiol.* **1980**, *100*, 147–160. [CrossRef]
35. Mockler, T.C.; Guo, H.; Yang, H.; Duong, H.; Lin, C. Antagonistic actions of Arabidopsis cryptochromes and phytochrome B in the regulation of floral induction. *Development* **1999**, *126*, 2073–2082. [CrossRef] [PubMed]
36. Ferrier, T.; Matus, J.T.; Jin, J.; Riechmann, J.L. *Arabidopsis* paves the way: Genomic and network analyses in crops. *Curr. Opin. Biotechnol.* **2011**, *22*, 260–270. [CrossRef] [PubMed]
37. Dere, S.; Gunes, T.; Sivaci, R. Spectrophotometric determination of chlorophyll—A, b and total carotenoid contents of some algae species using different solvents. *Turk. J. Bot.* **1998**, *22*, 13–17.
38. Tsukamoto, A.; Hirai, T.; Chin, D.P.; Mii, M.; Mizoguchi, T.; Mizuta, D.; Yoshida, H.; Olsen, J.E.; Ezura, H.; Fukuda, N. The FT-like gene PehFT in petunia responds to photoperiod and light quality but is not the main gene promoting light quality-associated flowering. *Plant Biotechnol.* **2016**, *33*, 297–307. [CrossRef]
39. Berenschot, A.S.; Quecini, V. A reverse genetics approach identifies novel mutants in light responses and anthocyanin metabolism in petunia. *Physiol. Mol. Biol. Plants* **2014**, *20*, 1–13. [CrossRef] [PubMed]

plants

Article

Effect of Various LED Light Qualities, Including Wide Red Spectrum-LED, on the Growth and Quality of Mini Red Romaine Lettuce (cv. Breen)

Joo Hwan Lee [1], Yong Beom Kwon [1], Yoo Han Roh [1], In-Lee Choi [2], Jidong Kim [3], Yongduk Kim [4], Hyuk Sung Yoon [5,*] and Ho-Min Kang [1,2,*]

1 Interdisciplinary Program in Smart Agriculture, Kangwon National University, Chuncheon 24341, Republic of Korea; ot2581@naver.com (J.H.L.); nm96727@naver.com (Y.B.K.); nuh3722@naver.com (Y.H.R.)
2 Agricultural and Life Science Research Institute, Kangwon National University, Chuncheon 24341, Republic of Korea; cil1012@kangwon.ac.kr
3 FutureGreen Co., Ltd., Yongin 17095, Republic of Korea; jidong.kim@futuregreen.co.kr
4 Cheorwon Plasma Research Institute, Cheorwon 24062, Republic of Korea; ydkim@cpri.re.kr
5 The Waksman Institute of Microbiology, Rutgers the State University of New Jersey, Piscataway, NJ 08854, USA
* Correspondence: yhs802@naver.com (H.S.Y.); hominkang@kangwon.ac.kr (H.-M.K.)

Citation: Lee, J.H.; Kwon, Y.B.; Roh, Y.H.; Choi, I.-L.; Kim, J.; Kim, Y.; Yoon, H.S.; Kang, H.-M. Effect of Various LED Light Qualities, Including Wide Red Spectrum-LED, on the Growth and Quality of Mini Red Romaine Lettuce (cv. Breen). *Plants* **2023**, *12*, 2056. https://doi.org/10.3390/plants12102056

Academic Editors: Hail Rihan and Mick Fuller

Received: 20 March 2023
Revised: 26 April 2023
Accepted: 17 May 2023
Published: 22 May 2023

Abstract: Recently, LEDs with various light qualities have been used in closed plant factories, and they are known to have different effects on the growth and quality of crops. Therefore, this study was conducted to investigate the change in growth and quality in mini red romaine lettuce using LEDs with various light qualities. Wide red spectrum (WRS)-LEDs, blue (B)-LEDs, blue + red (BR)-LEDs, red (R)-LEDs, and white (W)-LEDs were used as the artificial light sources. Regarding growth, the R-LED treatment showed the most positive effect, but the leaf shape was not normal and the Hunter b* value was not suitable because it was higher than that of the other treatments. The Hunter a*, SPAD, and NDVI values of the B- and BR-LED treatments were effective, but this was not the case for those of the R- and W-LED treatments. The anthocyanin reflectance index 1 (ARI1) was 20 times higher in the B-LED treatment than in the R-LED treatment, and the ascorbic acid content was the highest in the WRS-LED treatment. In the sensory evaluation, bitterness and sweetness showed opposite tendencies. Regarding the overall preference, the BR-LED treatment received the highest score. Correlation analysis showed that the bitterness was closely correlated with the anthocyanin content and leaf color. Taken together, BR-LEDs provided a good top fresh weight, dark red leaves, and high anthocyanin and ascorbic acid contents, with the highest overall preference; therefore, BR-LEDs were the most suitable for the cultivation of mini red romaine lettuce.

Keywords: led; light quality; mini red romaine lettuce; anthocyanin; bitterness

1. Introduction

According to the Food and Agriculture Organization, more than one million hectares of lettuce are cultivated worldwide, with a production of more than 22 million tons in 2022 [1]. Lettuce, which is mainly consumed as a fresh-cut salad, can provide phenolic compounds, flavonoids, carotenoids, and vitamin C, and is consumed worldwide, as it can be produced year-round [2]. In particular, red romaine lettuce, which has a red leaf color, can play a beneficial role in health by accumulating a large amount of anthocyanin, which is well known for its antioxidant action, in its tissues [3]. One of the major characteristics of lettuce is its bitter taste, and lactucopicrin, which is among the major bitter sesquiterpene lactones (BSLs), was reported to be the main cause [4]. In addition, BSLs can act as an important factor in consumer purchasing, and if the content is too high, preference can be significantly reduced [5,6].

Currently, LEDs (light-emitting diodes) are the main light source among artificial light sources used in plant factories [7]. LEDs have various advantages over existing artificial light sources, such as a long lifetime, higher electrical conversion efficiency, and lower heat generation and price [8,9]. However, the most important factor in crop cultivation is that they can produce numerous types of light quality (spectrum) by combining different wavelengths, providing the optimal light quality according to the crop type and growth stage [10,11]. It is reported that these different combinations of LEDs can have various effects on the growth and quality of crops [7,8].

In plants, the main photosynthetic pigments, namely, chlorophyll *a* and *b*, absorb most of the blue and red light in the range of 400–700 nm, which is known as photosynthetically active radiation (PAR). Additionally, they can respond through photoreceptors, such as cryptochrome, which detects blue light, and phytochrome, which detects red light [11]. Previous studies also reported that the combination of blue and red light is the most effective for the growth and development of many leafy vegetables, including lettuce [12]. Based on the above reasons, blue (B)-, red (R)-, blue + red (BR)-, white (W)-, and wide red spectrum (WRS)-LEDs were used as artificial light sources in this study.

Blue light emitted from B-LEDs (400–500 nm) is known to act as an important wavelength for the formation of biomass, anthocyanin, chloroplasts and chlorophyll, and photomorphogenesis in lettuce and various crops [13,14], but they have the disadvantages of producing a small leaf size and a slow growth rate when used alone at a high light intensity [15,16]. Conversely, R-LEDs (600–700 nm) act as the most effective wavelength for the growth rate and photosynthesis of lettuce in plant factories, but it reduces the phenol content and chlorophyll relative to B-LED [11,17].

Artificial light sources widely used as mixed light include BR-LEDs and W-LEDs. BR-LEDs (400–500 + 600–700 nm) were reported to improve the accumulation of phenolic compounds and growth through a synergistic effect when irradiated with a mixture of B- and R-LEDs compared with when B- and R-LEDs were used alone [18,19]. In addition, Kang et al. [15] reported that BR(2:8)-LEDs provided the greatest increase in the photosynthetic rate of lettuce compared with B- and R-LEDs alone. Unlike BR-LEDs, W-LEDs (400–700 nm) contain a large amount of green light in their spectrum. In a previous study, it was reported that the biomass and growth rate of lettuce increased when green light was added to a BR-LED [20,21]. However, it has also been found that the photosynthetic rate is greatly reduced without affecting the growth of lettuce [15].

WRS-LEDs are artificial light sources that use quantum dots, whose optical and electrical properties change when a semiconductor is reduced to nanometer (nm) size. They have a wider light distribution angle than conventional LEDs, and thus, their uniformity is high, and they have the advantage of being able to produce the Emerson synergy effect, which increases the photosynthesis rate compared with when other wavelengths are irradiated independently [22]. In this experiment, the wavelengths of WRS-LEDs included 26.5% blue (B) light (400–500 nm), 12.2% green (G) light (500–600 nm), 50.8% wide-red (R) light (600–700 nm), and 10.5% far-red (FR) light (700–800 nm). FR light has been mentioned as a necessity for plants to perform efficient photosynthesis and photochemistry [23], and it is sensed by phytochrome, together with red light, and is known to show higher leaf transmittance than red light [11]. In addition, it causes a shade avoidance reaction in plant growth, increases the size of leaves, and elongates stems, which can cause significant changes in plant morphology [24]. According to a previously reported study, when lettuce was treated by adding FR light to blue and red light, the biomass increased by 39% and 25%, respectively, and the appearance of the plant was changed to improve the light use efficiency [25]. In addition, Hwang et al. [26] reported that as a result of cultivating tomatoes, peppers, cucumbers, and watermelons by supplementing far-red rays with cool-white LEDs, the hypocotyl length and dry weight of seedlings increased as the light intensity of far-red rays increased. Furthermore, Tan et al. [27] reported far-red-induced changes in plant height, leaf structure and shape, stomatal response, chloroplast development, biomass, photosynthetic pigment and fluorescence, electron transport, carbon assimilation, etc., in various crops.

As explained in the papers referenced above, the growth and quality of plants can vary depending on the light quality, and among LEDs of various light qualities, BR-LED, which is a mixed light, was found to be effective in cultivation [8,15,18]. In particular, WRS-LED is expected to be more effective than existing artificial light sources by making up for the shortcomings of existing LEDs, utilizing a wide light distribution angle, a wide red spectrum, and far-red rays. Therefore, this study was conducted to identify the growth and quality changes in mini red romaine lettuce (cv. Breen) using LEDs with various light qualities in a closed plant factory-type chamber.

2. Results

During the entire cultivation period, lettuce grown under R-LEDs produced the longest leaf length compared with the other treatments, followed by the WRS-LEDs. The difference in leaf length of lettuce cultivated under R-LEDs and WRS-LEDs up to the 28th day showed statistical significance, with a difference of about 2 cm, but there was no difference from the 35th day (Figure 1A). The number of leaves showed a consistent trend until the 49th day, except for the 14th day of cultivation. At the end of cultivation, lettuce grown under R-LEDs produced the most number of leaves, with 51.7 leaves, showing a significant difference from the other treatments, followed by the BR-, WRS-, W-, and B-LED treatments in order. On the 49th day, lettuce grown under R-LEDs produced 35% more leaves than lettuce grown in B-LEDs, which produced the fewest leaves (Figure 1B).

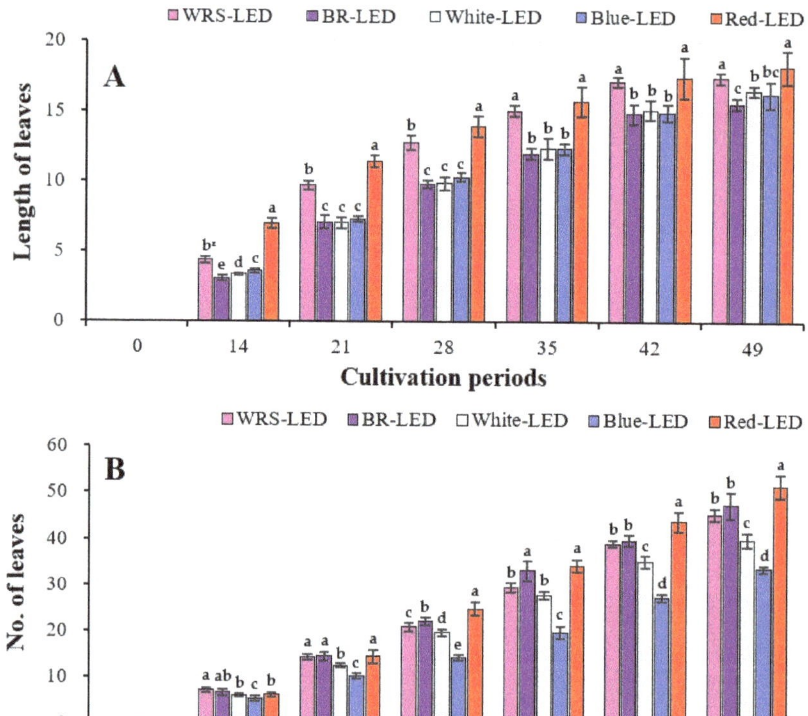

Figure 1. Change in the length of leaves (**A**) and the number of leaves (**B**) of mini red romaine lettuce cultivated under LEDs with various light qualities for 49 days. Vertical bars represent ± SD (n = 6–8). [z] Mean separation within columns by Duncan's multiple range test at 5% level. Values marked with different letters indicate significant differences according to Duncan's multiple range test at the 5% level.

At the end of cultivation, the top fresh weight was significantly the biggest in lettuce grown in the R-LED treatment among all treatments, producing the highest leaf length and the number of leaves. The top fresh weight of lettuce grown under the R-LED treatment on the 49th day was 25.7% bigger than that of the BR-LED treatment, which was the second biggest, and 43.5% bigger than that of the B-LED treatment, which was the lowest. In the R-LED treatment, which produced the biggest top fresh weight, the number of leaves and top fresh weight on day 49 showed the same trend. However, there was no statistical significance between the BR- and WRS-LEDs, or between the W- and B-LEDs (Table 1). The top dry matter ratio was the highest in lettuce cultivated under the BR-LED treatment, which produced the second-biggest top fresh weight, but there was no statistically significant difference from the rest of the treatments, except for the R-LED treatment. Lettuce grown in the R-LED treatment, which produced the biggest top fresh weight, showed the lowest top dry matter ratio, and conversely, the B-LED treatment, which produced the smallest top fresh weight, showed the second-highest top dry matter ratio (Table 1).

Table 1. Top fresh weight and top dry matter ratio of mini red romaine lettuce cultivated under LEDs with various light qualities on the final day.

Treatments	Top Fresh Weight (g)	Top Dry Matter Ratio (%)
WRS-LED	72.41 ± 2.97 b[z]	5.21 ± 0.33 a
BR-LED	74.47 ± 3.10 b	5.48 ± 0.39 a
W-LED	57.61 ± 4.18 c	5.03 ± 0.33 ab
B-LED	56.65 ± 5.77 c	5.35 ± 0.20 a
R-LED	100.18 ± 9.92 a	4.68 ± 0.22 b

[z] Means with different letters within column indicate statistically significant differences by Duncan's multiple range test at the 5% level.

As for chromaticity, Hunter L* (closer to 100, whiter; closer to 0, blacker) was the highest in the R-LED treatment, and lettuce grown under BR- and B-LEDs showed a lower value, resulting in a darker leaf color. Regarding Hunter a* (+ is redder and − is greener), lettuce grown in the B-LED treatment produced the deepest red color, and there was no significant difference from the BR-LED treatment, which had the second-highest value. Hunter a* values showed negative results with green leaves only in the R-LED treatment with the best growth, and all lettuce grown in other treatments produced red leaves, which can be visually confirmed in Figure 2. Regarding Hunter b* (+ is yellower and − is bluer), the B- and BR-LED treatments, which produced the deepest red color in lettuce leaves, had the lowest values without statistical significance. In contrast, lettuce grown in the R-LED treatment, which produced green leaves, showed a significantly higher value compared with the other treatments (Table 2).

Table 2. Leaf color of mini red romaine lettuce cultivated under LEDs with various light qualities on the final day.

Treatments	Hunter L*	Hunter a*	Hunter b*
WRS-LED	35.53 ± 2.06 c[z]	2.20 ± 1.32 b	9.68 ± 2.72 b
BR-LED	32.48 ± 0.80 d	4.37 ± 0.34 a	3.03 ± 1.23 c
W-LED	38.25 ± 1.37 b	1.12 ± 1.02 c	9.50 ± 2.47 b
B-LED	32.27 ± 0.48 d	4.92 ± 0.30 a	3.20 ± 0.58 c
R-LED	47.83 ± 1.40 a	−4.43 ± 0.54 d	26.93 ± 2.93 a

[z] Means with different letters within column indicate statistically significant differences by Duncan's multiple range test at the 5% level.

Figure 2. Mini red romaine lettuce cultivated for 49 days under LEDs with various light qualities.

The soil plant analysis development (SPAD), which can represent the chlorophyll content, was the highest in lettuce grown with the BR-LED treatment and was significantly higher than that of the R-LED treatment, which had the lowest chlorophyll content, by more than 24%. In addition, in lettuce cultivated in the BR-, B-, WRS-, W-, and R-LED treatments, the chlorophyll content showed the same trend as the top dry matter ratio, but the difference between the treatments was more pronounced in the chlorophyll content (Table 3). The normalized difference vegetation index (NDVI) is a value that is proportional to the chlorophyll change and plant health status. Similar to SPAD, it was the highest for lettuce grown under the B-LED and BR-LED treatments without significance, and the R-LED treatment showed the lowest value. In the Polypen manual for measuring the NDVI, the value ranges are stated as 0.5–0.9 for healthy leaves and 0.2–0.4 for unhealthy leaves. Only the lettuce cultivated under WRS-, BR-, and B-LED treatments had values corresponding to the range for healthy leaves, with values of 0.519, 0.530, and 0.550, showing statistical significance compared with the W- and R-LED treatments, which had values corresponding to the range for unhealthy leaves (Table 3). The anthocyanin reflectance index 1 (ARI1) reflects changes in the anthocyanin content. In this study, the trend of ARI1 was the same as that of Hunter a*. Lettuce grown under the B-LED treatment produced the highest anthocyanin content, which was more than 3 times higher than that of the BR-LED, WRS-LED, and W-LED treatment groups, and 20 times higher than that of the R-LED treatment group, which produced the lowest content. Summarizing the results of ARI1, it was shown that blue light increased the red color expression of red romaine lettuce, while red light and green light decreased it. In addition, in the case of Hunter a*, there was no significant difference between lettuce cultivated under the B- and BR-LED treatments, indicating that blue light was responded to more sensitively than red light to produce the red color expression of red romaine lettuce. However, ARI1 showed more than twice the difference between the B- and BR-LED treatments (Table 3). It seems that the degree of difference in the results between the treatment groups was due to the difference in the measurement method of Hunter a* and ARI1. The ascorbic acid content was the highest in lettuce grown under the WRS-LED treatment at 4.40 mg/100 g FW, which was 38% significantly higher than that under the R-LED treatment, which had the lowest value. The reason why the lettuce grown under the WRS-LED treatment was able to produce the highest ascorbic acid content is thought to be due to the FR light. However, it did not seem to have a significant effect, as there was no significant difference between the BR- and B-LED treatments. The red light treatment demonstrated a low ascorbic acid synthesis ability when used alone in lettuce, but when irradiated with blue light, the content was higher than that of blue light alone due to the positive synergy (Table 3). Under the W-LED treatment, which produced the lowest content among the mixed lights, the green light in the spectrum seemed to interfere with the ability of the blue light to synthesize ascorbic acid.

Table 3. SPAD, NDVI, ARI1, and ascorbic acid of mini red romaine lettuce cultivated under LEDs with various light qualities on the final day.

Treatments	SPAD	NDVI	ARI1	Ascorbic Acid (mg/100 g FW)
WRS-LED	31.0 ± 0.52 c[z]	0.519 ± 0.02 b	0.754 ± 0.18 b	4.40 ± 0.23 a
BR-LED	38.8 ± 0.68 a	0.530 ± 0.01 ab	0.809 ± 0.29 b	4.15 ± 0.73 a
W-LED	29.5 ± 0.93 d	0.471 ± 0.03 c	0.473 ± 0.03 b	3.39 ± 0.17 b
B-LED	34.9 ± 0.47 b	0.550 ± 0.02 a	1.951 ± 0.52 a	4.03 ± 0.14 a
R-LED	29.3 ± 2.16 d	0.441 ± 0.03 d	0.125 ± 0.03 c	2.73 ± 0.38 c

[z] Means with different letters within column indicate statistically significant differences by Duncan's multiple range test at the 5% level.

The sweetness was the highest for lettuce grown under the R-LED treatment and the lowest for the B-LED treatment. Contrary to sweetness, bitterness was the highest for lettuce cultivated under B-LEDs and showed a stronger bitter taste than lettuce cultivated under R-LEDs, which was the lowest, by 53%. In addition, there was a clear difference between the treatment groups in bitterness rather than sweetness. The results of sweetness and bitterness were related to the Hunter a* and ARI1 results: sweetness was low and bitterness was high when the leaf color was deep red, and the opposite tendency was shown when the leaf color was green. However, it is known that a strong bitter taste can reduce consumers' purchase preferences [5,6]; in this study, the overall preference for B-LED-treated lettuce, which had a red leaf color and the strongest bitter taste, was the second lowest among all treatments groups (Figure 3). Sourness was the highest for lettuce grown under BR-LEDs, but due to the nature of these crops, the sourness was investigated as low, i.e., less than 1–2 points, in all treatment groups; therefore, it did not seem to be affected by the light quality. Among the sensory evaluation items, the difference in leaf color was the greatest between the treatment groups, and the change due to the light quality was large. As for leaf color, which tended to show the same trend as Hunter a* and ARI1, the B-LED and BR-LED treatments, which produced dark red leaves, scored high, followed by the WRS-LED, W-LED, and R-LED treatments. The highest flavor score was obtained by lettuce cultivated under the BR-LED treatment, while the lowest was found for the W-LED treatment. However, since there were no similar or identical trends among the survey items investigated in this study, additional research on the flavor of lettuce according to the light quality is necessary. Texture obtained the highest preference score for lettuce grown under R-LEDs. The reason for this is not indicated in Figure 3, but the judges said that the leaf tissue was soft, and thus, gave it a higher score than the other treatments. Finally, for the overall preference, lettuce grown under the BR-LED treatment received the highest score, followed by the W-LED, WRS-LED, B-LED, and R-LED treatments, but there was no statistical significance (Figure 3).

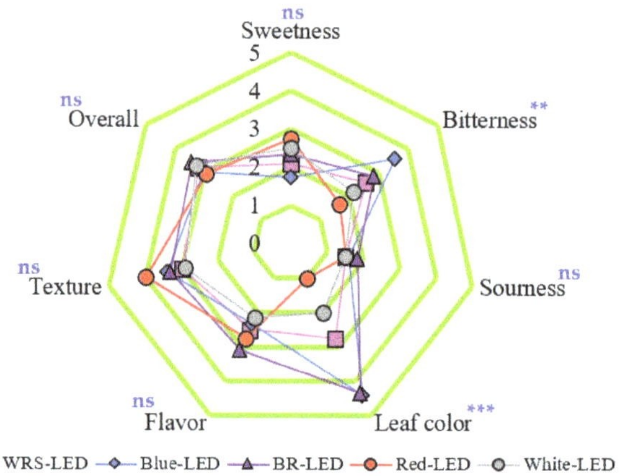

Figure 3. Sensory evaluation of mini red romaine lettuce cultivated under LEDs with various light qualities on the final day. ns, **, and *** indicate non-significant and significant differences at $p < 0.01$, and 0.001, respectively. Sweetness, bitterness, sourness, and flavor: 5 = very strong, 4 = strong, 3 = normal, 2 = faint, and 1 = very faint. Leaf color: 5 = very deep, 4 = deep, 3 = normal, 2 = light, and 1 = very light. Texture and overall preference: 5 = very good, 4 = good, 3 = normal, 2 = bad, and 1 = very bad.

The growth and quality characteristics of mini red romaine lettuce were analyzed via correlation analysis, as shown in Figure 4. First, the top fresh weight was found to have a significant positive correlation with the number of leaves, with 0.899, indicating that the increase was due to the number of leaves rather than the length of the leaves. Hunter a* showed a high negative correlation with Hunter b* at the $p < 0.01$ level, and it increased statistically significantly with bitterness in the sensory evaluation. In addition, Hunter a*, the top dry matter ratio, the ascorbic acid content, and the bitterness had important effects on the NDVI, which can be used as an indicator of plant health. ARI1, which is proportional to the anthocyanin content, showed a negative correlation with sweetness and a positive correlation with bitterness, while sweetness and bitterness showed a negative correlation with each other at the 95% level. As a result, in this study, the plants with a dark red color, high top dry matter ratio, high ascorbic acid content, and strong bitter taste were healthy, whereas dark yellow leaves were unhealthy. Comparing the above results with the various LEDs used in this study, it can be said that lettuce grown under BR- and B-LEDs was healthy, but lettuce grown under R-LEDs was not. WRS-LEDs also produced most of the conditions for healthy lettuce, but could not produce all of them due to a low expression of leaf color. In addition, the sweetness and bitterness of the lettuce showed opposite tendencies, where the higher the anthocyanin content, the stronger the bitterness and the lower the sweetness (Figure 4).

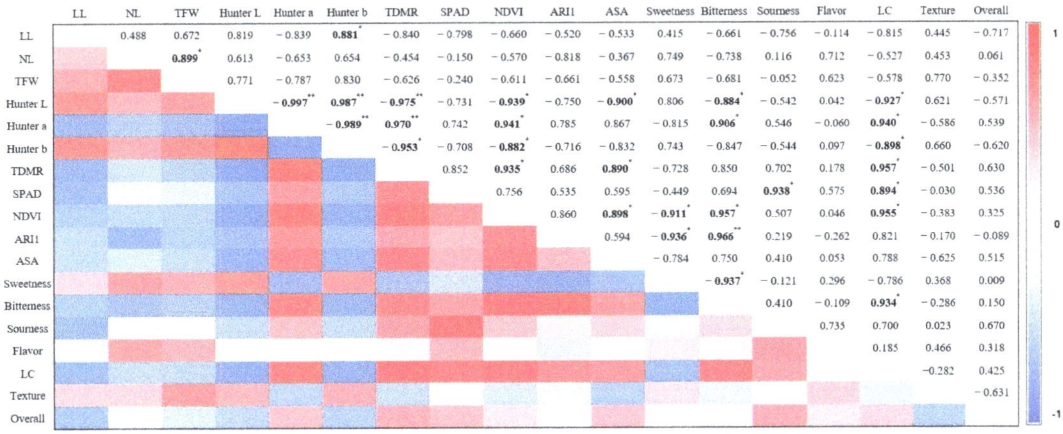

Figure 4. Correlation analysis of red romaine lettuce growth and quality characteristics when cultivated under LEDs with various light qualities on the final day. LL: leaf length; NL: number of leaves; TFW: top fresh weight; TDMR: top dry matter ratio; NDVI: normalized vegetation index; ARI1: anthocyanin reflectance index 1; ASA: ascorbic acid; LC: leaf color. * indicates a significant correlation at $p < 0.05$; ** indicates an extremely significant correlation at $p < 0.01$.

3. Discussion

It is known that LEDs, which can be easily controlled according to the requirements of plants and can have various light qualities, can affect plant growth, biomass, and functional compounds in various ways [7,8].

Shimizu et al. [28] suggested that red light may be the most effective wavelength for photosynthesis and growth rates when growing lettuce in a plant factory. In addition, in a previous study, the number of leaves and the photosynthetic rate also showed the highest values under red light, followed by a BR-LED treatment [29]. This is the same trend as that found in the results of this study, which provides a basis for the results of the longest leaf length and the largest number of leaves found in lettuce grown under red light alone (Figure 1A,B).

The 49th-day top fresh weight also showed the highest result under the R-LED treatment (Table 1). In a previous study, lettuce grown under red light also showed the biggest top fresh weight, which was attributed to the high photosynthetic rate [28]. However, Chen et al. [30] reported that lettuce grown under red light grew rapidly, but when the ratio of red light was over 70%, the petiole distortion was evident, and with 100% red light, the original lettuce shape was lost. Similarly, in this study, it was difficult to see that the lettuce cultivated under R-LEDs was commercially viable, as it showed heterogeneity in the leaves and overall shape compared with lettuce grown under other light qualities (Figure 2). Therefore, it seems that lettuce cultivation under BR-LEDs, which produced the second biggest top fresh weight after R-LEDs, as well as a normal leaf shape, is more suitable. It has been reported that among LEDs of various light qualities used in this study, FR light, which is included only in WRS-LEDs, can significantly change photomorphogenesis by causing shade avoidance symptoms during plant growth [24]. With these characteristics, it has been stated that adding far infrared rays to the existing light source can reduce seedling size variation within the same cultivation bed [31]. For lettuce grown in the WRS-LED treatment, the standard deviation of the top fresh weights by wider light distribution angle, wide red spectrum, and FR rays was the smallest at 2.94, confirming that cultivation among individuals within the same treatment group proceeded uniformly (Table 1). However, as there was no difference from lettuce cultivated under the BR-LED treatment in terms of the top fresh weight, the wider light distribution angle, wide red spectrum, and FR rays seem to have produced a greater uniform cultivation and shade avoidance effect than growth enhancement. Lettuce grown under the B-LED and W-LED treatments showed poor growth compared with the other treatments (Table 1). This coincides with the results of Kang et al. [15], who found that blue light slowed down the growth rate of lettuce and that the green light comprising 30% of the W-LED did not have a positive effect on the growth rate. Previous papers reported that green light reduces the photosynthetic rate by reducing the chlorophyll content and stomatal conductance [32,33], but does not affect plant growth [15]. However, this does not mean that it does not affect lettuce growth at all, and it is thought that green light does not cause a direct growth reduction mechanism in lettuce, but indirectly affects the photosynthetic rate and stomatal conductance, leading to a growth reduction.

The degree of redness (Hunter a*) expressed in red romaine lettuce leaves is known to have a strong positive correlation with the actual anthocyanin content, with an R^2 value of 0.80 [34]. Blue light, for which the Hunter a* value was the highest in this study (Table 2), is particularly used as an essential wavelength for anthocyanin synthesis in red lettuce [13]. However, it has been reported that R-, G-, and FR-LEDs do not synthesize anthocyanin in lettuce or suppress the effect of blue light to prevent the red expression of the leaves [13,35,36], and the same result was confirmed in this study (Table 2). In a previous study, R-LEDs with a high ratio of red light and R:FR did not detect that a leaf was under other leaves, and thus, the expression of SAG (senescence-associated gene) family genes (e.g., x SAG13) related to leaf senescence was suppressed [37]. However, in this study, the uniquely high Hunter b* value in lettuce grown under the R-LED treatment were significantly negatively correlated with the NDVI (Figure 4). This means that the higher the Hunter b* result, the poorer the health of the plant, and the increase in yellow color in lettuce, which usually has green or red leaves, mainly means yellowing of the leaves. Therefore, in this study, it seems that the yellowing of the leaves progressed relatively quickly in lettuce cultivated under the R-LED treatment compared with the other treatments.

BR-LEDs are effective in promoting plant growth and biomass accumulation [8,15]. However, in this study, due to the characteristics of lettuce, there was no difference between the treatment groups, except for the R-LED group (Table 1), because the body water content was more than 95%. It was reported that FR light absorbs more water than BR light and increases the amount of water in the cell, which increases the expandability of each cell, thereby increasing the ratio of top fresh weight to top dry weight [38]. In addition, in previous studies, it was found that red light increased the top fresh weight but decreased the top dry weight [29] and that the top dry weight of red lettuce grown under R-LEDs was lower than that of lettuce grown under B-LEDs and BR-LEDs [35].

Blue light promoted 5-aminolevulinic acid (ALA), a precursor of chlorophyll tetrapyrrole, and suppressed the decrease in ALA caused by red light, resulting in the recovery of the chlorophyll concentration [39,40]. Zheng et al. [41] reported that lettuce grown in BR-LED, a mixed light, had a higher chlorophyll content than B- and R-LEDs alone due to a synergistic effect. Additionally, when irradiation was performed by adding FR light (50 $\mu mol \cdot m^{-2} \cdot s^{-1}$) to a BR-LED (200 $\mu mol \cdot m^{-2} \cdot s^{-1}$) for 16 h during the day and 1 h at the end of the day, the chlorophyll content of lettuce decreased [25]. Green light also downregulates transcription factors for chloroplast formation, reducing the chlorophyll content [42].

The NDVI is an indicator that can be used to check the health status, stress level, and chlorophyll concentration of plants by comparing the amount of red light, which is a part of visible light, with the amount of deflected NIR light. Alsina et al. [43] found the highest levels of chlorophyll $a + b$ when 'Lollo Bionda' lettuce was grown under blue light, followed by BR light, with the lowest levels under R light. In addition, during lemon balm cultivation, W-LEDs containing some green light had a negative effect on the NDVI [44]. Both NDVI and SPAD are used as indicators of chlorophyll content, and in this study, the two trends were similar, but the difference in the degree of significant difference between treatments was considered to be due to the different measured infrared wavelength bands (Table 3). SPAD's infrared wavelength band is measured at 940 nm, while that of the NDVI is measured in the 770–900 nm range.

The expression of genes that induce anthocyanin synthesis is induced by blue light and is known to be mediated by cryptochrome (Cry1) [45]. Cryptochrome, which absorbs and reacts with blue light treatment, has a total of two maximum absorptions at the 375 nm chromophore 5,10-methenyltetrahydrofolic acid (MTHF) and the 450 nm flavin adenine dinucleotide (FAD) chromophore flavin [46]. These flavin chromophores are reduced to half-forms under blue light, and cryptochromes become inactive under green and yellow light [42]. Light that can interfere with anthocyanin expression includes green light, red light, and far-red light. First, green light has an opposite tendency to blue light. In a previous study, when green and blue lights were used at the same time compared with when blue light was used alone, it was reported that in lettuce the oxidized flavin content was greatly reduced and the anthocyanin level was low [36]. In addition, it was reported that the completely reduced form of flavin showed the same movement as the flavin light balance in vivo caused by green light, and finally, the overall oxidized and reduced form of Cry1 was reduced [36]. In the case of red light, this can be explained by its effect on gene expression in the anthocyanin biosynthesis process. In previous studies, the gnl | UG | Lsa#S56341499 gene among lucoanthocyanidin oxidase (LDOX) coding genes and gnl | UG | Lsa#S58677322 gene among dihydroflavonol 4-reductase (DFR) coding genes were most frequently expressed under blue light during anthocyanin biosynthesis; however, they were not well expressed in lettuce grown under red light [37]. Finally, it was reported that in lettuce the anthocyanin concentration was reduced by up to 40% when FR light was also irradiated with white fluorescent light [13].

The assimilation of ascorbic acid in plants is significantly affected by light and temperature, and it is known that the light environment in particular has an important effect on the ascorbic acid metabolic pathway [47,48]. Chen et al. [49] reported that the ascorbic acid content in lettuce was higher when grown under B-LEDs and BR-LEDs compared with lettuce cultivated under R-LEDs. Adding green light to BR-LEDs resulted in a 44% reduced ascorbic acid content of lettuce compared with the use of BR-LEDs alone [50]. In addition, it was reported that W-LEDs supplemented with FR light increased the accumulation of ascorbic acid in green lettuce by 45% compared with W-LEDs alone, but they reduced the pigments and biomass [51]. However, as a result of adding red light to W-LEDs, there was no difference in the ascorbic acid content of lettuce compared with W-LEDs alone [51].

Even in recent studies, it is difficult to find content that sensory evaluation results for lettuce play an important role in deriving the final result, and most of them are comparisons of sensory evaluation parameters according to treatment groups [52–54]. In this study, there were some clearly distinguished results, such as bitterness and leaf color, but most did not show significant differences according to light quality. Therefore, this author believes that consumers' sensory evaluation may not play a large role in determining the optimal light quality, but can be used as a reference. However, consumers' sensory evaluation of lettuce can explain the difference for each parameter via comparison according to light quality. Meng et al. [54] reported that $B_{20}G_{60}R_{100}$ had significantly higher sweetness than $B_{100}G_{60}R_{20}$ as a result of the sensory evaluation of 'Rouxai' red oakleaf lettuce. Green light supplemented with oak lettuce was reported to be related to the activity or synthesis of enzymes related to sugar metabolism [51]. In this study, W-LEDs containing green light produced a higher sugar content than lettuce grown under BR-LEDs (Figure 3), but Nur Syafini et al. [55] reported that soluble solid contents (°Brix) were significantly reduced when green light was added to BR-LEDs. These conflicting results may be due to differences in the respective ratios of red, blue, and green light within the light source, or because the sweetness perceived by humans and the measured soluble solid contents (°Brix) are not proportional. It is known that the bitterness of lettuce is mainly affected by the content of the compounds lactucin, 8-deoxylactucin, and lactucopicrin, which are types of bitter sesquiterpene lactones (BSLs) [4]. In a previous study, the bitterness of lettuce showed the lowest result with the use of R-LEDs alone, and when $B_{20}R_{160}$ and $B_{100}R_{80}$ were compared, bitterness was found to increase in $B_{100}R_{80}$, indicating that blue light increases bitterness [54]. This suggests the possibility that biosynthetic enzymes may be affected by the light quality during the biosynthesis of BSLs. However, there are currently very few studies that clearly report the degree of bitterness and the BSL content of lettuce according to various light qualities. In the case of previous studies on lettuce sourness, a comparison experiment was conducted as part of the sourness sensory evaluation according to the lettuce varieties [56], and an analysis was carried out on the content of acetic acid and lactic acid representing the sour taste in the packaging during the storage of fresh lettuce [57]. However, when compared with other important parameters, sourness, among the taste components in lettuce, was not considered very important, and there was no clear difference; therefore, it was difficult to find accurate information on comparative sourness studies of single varieties of lettuce according to various light qualities. In this study, flavor averaged around 2.5 points, with no significant difference between the treatment groups. Flavor and overall preference also showed no significant differences according to the light quality of the artificial light source [54]. In this study, the texture of lettuce grown under R-LEDs received the best evaluation (Figure 3), and Meng et al. [54] also reported that the highest score was obtained for lettuce grown under R-LEDs alone, followed by BR-LEDs and W-LEDs. In the literature, among the lettuce sensory evaluations according to the use of various LEDs, most of the results of food taste surveys were analyzed for correlations between parameters or the statistical significance of the results until recently, and the biochemical content was rarely mentioned. In addition, content sensory evaluations of lettuce grown under FR-LEDs are rare, and thus, further research should be conducted.

In this study, there was no significant correlation between Hunter a* and ARI1 values, with 0.785 (Figure 4), but this value was similar to the value found in a previous study showing a significant positive correlation, with 0.80 [34]. In addition, ARI1 showed a high positive correlation of 99% with bitterness (Figure 4), which can be attributed to the role of anthocyanins in the bitter taste as the leaf color becomes red. These results suggest that the changes in the anthocyanin content were directly related to bitterness. In a previous study, it was reported that malvidin-3-glucoside, which is an anthocyanin, activates the TAS2R7 receptor among the bitter taste receptors (TAS2R) in humans, resulting in a bitter taste [58]. In addition, it was shown that the light quality that increases the BSL content and the light quality that increases the anthocyanin content may be in the same wavelength range, and thus, additional research is needed. So far, studies on the bitter taste of lettuce have been limited to studies on the difference in the BSL content according to leaf color and cultivar [4], and there are no studies related to light factors, such as the light intensity, photoperiod, and light quality. Therefore, although this study did not investigate the BSL content, it was the first time the degree of bitterness of single cultivar red romaine lettuce according to various light qualities and the relationship between bitterness and anthocyanin content were mentioned. In a previous paper, the total BSL content of red lettuce was significantly higher than that of the green variety, but there was no consistent trend between the total BSL content and sugar content (°Brix) according to the leaf color [4].

4. Materials and Methods

4.1. Plant Material and Growing Conditions

Mini red romaine lettuce (cv. Breen, Johnny's Selected Seeds) was used as the material for testing. The temperature and humidity of the closed plant factory-type lettuce cultivation room were controlled at 20 ± 3 °C and 70 ± 5%RH, and the internal CO_2 concentration was 577 ± 67 ppm without any additional control. Lettuce was planted on a floating platform (50 × 350 × 490 mm) with 40 holes (5 × 8) of 33 mm diameter at a planting interval of 30 mm in a growing tray (130 × 400 × 540 mm). It was cultivated for a total of 49 days (7 weeks) at 7-day intervals using a deep-water culture method in which oxygen was supplied to the water using an aeration pump (SH-A2, Amazonpet, Daejeon, Republic of Korea). The experiment was conducted once, and the above hydroponic cultivation system was installed on a 3-tier shelf for growing. On day 0, 40 individuals were planted under the LEDs for each light quality, and survey items were investigated using 6 individuals on the 14th, 21st, and 28th days; 7 individuals on the 35th and 42nd days; and 8 individuals on the 49th day. As the investigation progressed, empty holes formed by the used lettuce were blocked with a sponge to prevent light from entering the nutrient solution, and the planting distance was increased by using the empty hole to move the seat as the lettuce grew.

4.2. Nutrition

The nutrition regime included Yamazaki lettuce nutrient solution and was divided into nutrition formulations A and B. Nutrition formulation A was composed of 472 mg·L^{-1} $Ca(NO_3)_2 \cdot 4H_2O$, 404 mg·L^{-1} KNO_3, and 48 mg·L^{-1} EDTA-NaFe(III). Nutrition formulation B was composed of 404 mg L^{-1} KNO_3, 115 mg L^{-1} $NH_4H_2PO_4$, 246 mg L^{-1} $MgSO_4 \cdot 7H_2O$, 6 mg L^{-1} H_3BO_3, 4 mg L^{-1} $MnSO_4 \cdot 5H_2O$, 0.4 mg L^{-1} $ZnSO_4 \cdot 7H_2O$, 0.1 mg·L^{-1} $CuSO_4 \cdot 5H_2O$, and 0.04 mg·L^{-1} $NO_2MoO_4 \cdot 2H_2O$. Nutrition formulations A and B prepared with the above compositions were used after being diluted 200 times. The pH was set to 6.0 $\pm$ 0.5, and the EC was supplied at 0.3 dS·cm^{-1} from the emergence of true leaves and 1.5 dS·cm^{-1} at the end of cultivation.

4.3. Light Treatments

The light-emitting diodes (LEDs) (40 W) were bar-type (20 $\times$ 30 $\times$ 1200 mm) wide red spectrum (WRS)-LEDs (400–800 nm) (Cheorwon Plasma Research Institute, Gangwon-do, Republic of Korea), blue + red (BR)-LEDs (400–500 nm + 600–700 nm) (HT402-1; BISSOL LED, Seoul, Republic of Korea), white (W)-LEDs (400–700 nm) (HT400-5700; BISSOL LED, Seoul, Republic of Korea), blue (B)-LEDs (400–500 nm) (HT400-Blue; BISSOL LED, Seoul, Republic of Korea), and red (R)-LEDs (600–700 nm) (HT400-Red; BISSOL LED, Seoul, Republic of Korea) (Figures 5 and 6). Each artificial light source was installed on a shelf on the 3rd floor, with 3 per floor, where the LED installation interval was 15 cm and the distance between the LEDs and the floating platform was 25 cm. As the growth progressed, the light intensity was adjusted with a light intensity controller (LED dimmer 20A; ZERO, Daejeon, Republic of Korea) and set to 200 $\pm$ 50 μmol·m^{-2}·s^{-1} using a Quantum radiometric probe (LP471PAR, Delta OHM, Veneto, Italy) in the dark condition, and the light and dark cycle was set to 16/8 h.

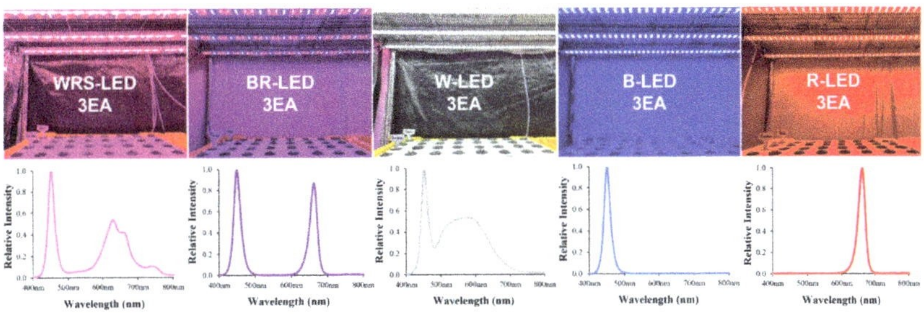

Figure 5. The cultivation environments of LEDs with various light qualities and the spectra of the WRS-LEDs, blue + red-LEDs, white-LEDs, blue-LEDs, and red-LEDs used in the experiment.

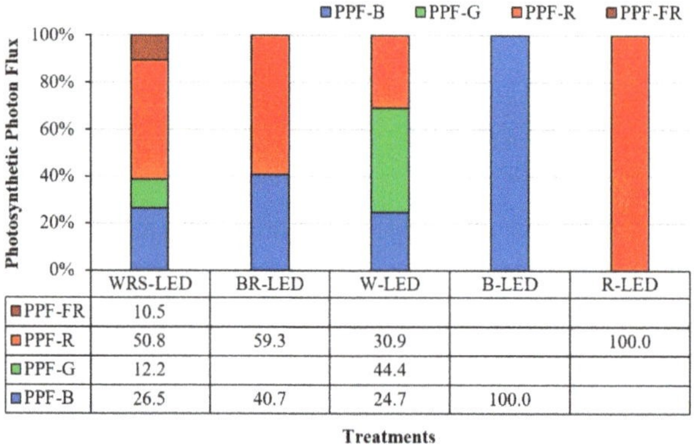

	WRS-LED	BR-LED	W-LED	B-LED	R-LED
PPF-FR	10.5				
PPF-R	50.8	59.3	30.9		100.0
PPF-G	12.2		44.4		
PPF-B	26.5	40.7	24.7	100.0	

Figure 6. PPF-B (400~499 nm), PPF-G (500~599 nm), PPF-R (600~699 nm), and PPF-NIR (700–780 nm) 100% stacked bar graph for the LED fixtures used in the experiment.

4.4. Change in Growth

The length of leaves and the number of leaves were measured every 7 days from the 14th day after planting using 6–8 plants, and the leaf length was measured using an electronic vernier caliper, while the leaf number was counted directly. As for the change in growth at the end of cultivation, the top fresh weight and top dry matter ratio were investigated using 8 plants. The top fresh weight was measured using an electronic scale (PB602-S, Mettler Toledo, Switzerland), and the top dry matter ratio was calculated using the formula shown below after drying at 80 °C for 72 h:

$$\text{Top dry matter ratio} = (\text{Dry weight}/\text{Fresh weight}) \times 100$$

4.5. Change in Quality

The leaf color (Hunter L*, a*, b*), soil plant analysis development (SPAD), normalized vegetation index (NDVI), anthocyanin reflectance index 1 (ARI1), and ascorbic acid content were investigated using 8 plants at the end of cultivation. The leaf color (Hunter L*, a*, b*) was measured with a Chroma Meter (CR-400, Konica Minolta Sensing, Inc., Japan), and the chlorophyll content was measured using a chlorophyll meter (SPAD-502 plus, Konica Minolta, Japan). The NDVI and ARI1 were measured with a polypen RP410 (Photon System Instruments Ltd., Drásov, Czech Republic), and the results were calculated according to the following equation:

$$\text{NDVI} = (\text{NIR} - \text{Red})/(\text{NIR} + \text{Red})$$

$$\text{ARI1} = (\text{R550})^{-1} - (\text{R700})^{-1}$$

The ascorbic acid content was determined according to the method of Arvanitoyannis et al. [59]. An amount of 18 mL of distilled water was added to 2 g of the sample, homogenized for 90 s with a homogenizer (HZ1, LABTron Co., Ltd., Bucheon, Republic of Korea), and then centrifuged with a centrifugal separator (Mega 17R, Hanil Science Industrial Co., Incheon, Republic of Korea). Using the supernatant obtained after centrifugation, the ascorbic acid content was measured with an RQ flex reflectometer (Merck RQ flex 2, Darmstadt, Germany).

4.6. Sensory Evaluation

Sensory evaluation according to the light quality of the artificial light sources was performed according to the quantitative descriptive analysis (QDA) method, as outlined in the procedures included in the standard Sensory Profiling ISO 13299:2016 [60] used in Matysiak et al. [52]. For the sensory evaluation, on the date of cultivation completion, 15 judges who had experience in the sensory evaluation of vegetables were surveyed on the sugar content, bitterness, sourness, flavor, leaf color, texture, and overall preference. The sensory evaluation score was set in units of 1 point, ranging from 1 to 5 points. Sweetness, bitterness, sourness, and flavor were given a score closer to 5 points if the specific taste and aroma were stronger. As for the leaf color, as the object of study was red romaine lettuce, the deeper the red color, the more points were given. Texture and overall preference were evaluated according to the subjective tendencies of the judges, where the higher the satisfaction, the more points were given.

4.7. Statistical Analysis

For statistical analysis, data statistical characteristics (correlation analysis, principal component analysis) were confirmed using the Microsoft Excel 2016 program and IBM SPSS Statistics version 26 program. The data were evaluated via ANOVA (analysis of variance), and the comparison of differences between the averages of the investigation items of the treatment groups was analyzed at the $p < 0.05$ level using Duncan's multiple range test. The standard deviation (SD) of each mean is indicated.

5. Conclusions

Summarizing the above results, the R-LED treatment was the best in terms of growth (leaf length, number of leaves, top fresh weight) according to the LEDs with various light qualities, but it was unsuitable since although it was red lettuce, the leaf color was not red, and the chlorophyll content, NDVI, and ascorbic acid content were the lowest. Conversely, the B-LED treatment produced good quality lettuce but with very low growth; therefore, it was not suitable for lettuce cultivation. WRS-LED was expected to show the best growth and quality change compared with other LEDs used in this study due to its wide light distribution angle, wide red spectrum, and FR light. However, compared with BR-LED, which received a positive evaluation in terms of growth and quality among existing LEDs with various light qualities, there was no noticeable advantage other than equal cultivation within the same treatment group. In this study, the BR-LED treatment produced a great top fresh weight, which is considered important in lettuce cultivation, along with the R-LED treatment. In addition, the value of Hunter a*, which is a measure of leaf redness, and ARI1, which reflects the anthocyanin content, were the highest after the B-LED treatment. The top dry weight ratio, SPAD, and overall preference showed the highest results among all the treatment groups, and ascorbic acid, which acts as an antioxidant, also had the second-highest content. Therefore, the BR-LED treatment was the most suitable for growing mini red romaine lettuce (cv. Breen) in a closed chamber.

Author Contributions: Conceptualization, J.H.L., I.-L.C. and H.-M.K.; methodology, J.H.L., I.-L.C. and H.-M.K.; formal analysis, J.H.L., Y.B.K. and I.-L.C.; investigation, J.H.L., Y.B.K. and Y.H.R.; resources, J.K., Y.K. and H.-M.K.; data curation, J.H.L., Y.B.K., Y.H.R., I.-L.C. and H.-M.K.; writing—original draft preparation, J.H.L.; writing—review and editing, J.H.L., I.-L.C., H.S.Y. and H.-M.K.; visualization, J.H.L.; supervision, I.-L.C. and H.-M.K.; project administration, J.H.L.; funding acquisition, I.-L.C. and H.-M.K. All authors have read and agreed to the published version of the manuscript.

Funding: This work was supported by the Korea Institute of Planning and Evaluation for Technology in Food, Agriculture and Forestry (IPET) through the Technology Commercialization Support Program, which was funded by the Ministry of Agriculture, Food and Rural Affairs (MAFRA) (122056-3), and the Republic of Korea and the Basic Science Research Program through the National Research Foundation of Korea (NRF) funded by the Ministry of Education (NRF-2021R1A6A1A03044242).

Data Availability Statement: Data is contained within the article.

Acknowledgments: The authors express special thanks to the professor and laboratory personnel for their aid during this experiment.

Conflicts of Interest: The authors declare no conflict of interest.

References

1. FAOSTAT. Available online: https://www.fao.org/faostat/en/#data (accessed on 17 November 2021).
2. Kim, M.J.; Moon, Y.; Kopsell, D.A.; Park, S.; Tou, J.C.; Waterland, N.L. Nutritional value of crisphead 'Iceberg' and romaine lettuces (*Lactuca sativa* L.). *J. Agric. Sci.* **2016**, *8*, 6. [CrossRef]
3. Islam, M.Z.; Lee, Y.T.; Mele, M.A.; Choi, I.L.; Kang, H.M. The effect of phosphorus and root zone temperature on anthocyanin of red romaine lettuce. *Agronomy* **2019**, *9*, 47. [CrossRef]
4. Seo, M.W.; Yang, D.S.; Kays, S.J.; Lee, G.P.; Park, K.W. Sesquiterpene lactones and bitterness in Korean leaf lettuce cultivars. *HortScience* **2009**, *44*, 246–249. [CrossRef]
5. Park, K.W.; Lee, J.M. Wrap-ups: A unique method of vegetable usage and consumption in Korea. *Chronica Hortic.* **2006**, *46*, 13–15.
6. Drewnowski, A.; Gomez-Carneros, C. Bitter taste, phytonutrients, and the consumer: A review. *Am. J. Clin. Nutr.* **2000**, *72*, 1424–1435. [CrossRef]
7. Nicole, C.C.S.; Charalambous, F.; Martinakos, S.; Van De Voort, S.; Li, Z.; Verhoog, M.; Krijn, M. Lettuce growth and quality optimization in a plant factory. *VIII Int. Symp. Light Hortic.* **2016**, *1134*, 231–238. [CrossRef]
8. Bantis, F.; Smirnakou, S.; Ouzounis, T.; Koukounaras, A.; Ntagkas, N.; Radoglou, K. Current status and recent achievements in the field of horticulture with the use of light-emitting diodes (LEDs). *Sci. Hortic.* **2018**, *235*, 437–451. [CrossRef]
9. Gomez, C.; Morrow, R.C.; Bourget, C.M.; Massa, G.D.; Mitchell, C.A. Comparison of intracanopy light-emitting diode towers and overhead high-pressure sodium lamps for supplemental lighting of greenhouse-grown tomatoes. *HortTechnology* **2013**, *23*, 93–98. [CrossRef]
10. Singh, D.; Basu, C.; Meinhardt-Wollweber, M.; Roth, B. LEDs for energy efficient greenhouse lighting. *Renew. Sustain. Energy Rev.* **2015**, *49*, 139–147. [CrossRef]
11. Ahmed, H.A.; Tong, Y.-X.; Yang, Q.-C. Optimal control of environmental conditions affecting lettuce plant growth in a controlled environment with artificial lighting: A review. *S. Afr. J. Bot.* **2020**, *130*, 75–89. [CrossRef]
12. Piovene, C.; Orsini, F.; Bosi, S.; Sanoubar, R.; Bregola, V.; Dinelli, G.; Gianquinto, G. Optimal red: Blue ratio in led lighting for nutraceutical indoor horticulture. *Sci. Hortic.* **2015**, *193*, 202–208. [CrossRef]
13. Li, Q.; Kubota, C. Effects of supplemental light quality on growth and phytochemicals of baby leaf lettuce. *Environ. Exp. Bot.* **2009**, *67*, 59–64. [CrossRef]
14. Liu, Y.; Schouten, R.E.; Tikunov, Y.; Liu, X.; Visser, R.G.; Tan, F.; Bovy, A.; Marcelis, L.F. Blue light increases anthocyanin content and delays fruit ripening in purple pepper fruit. *Postharvest Biol. Technol.* **2022**, *192*, 112024. [CrossRef]
15. Kang, W.H.; Park, J.S.; Park, K.S.; Son, J.E. Leaf photosynthetic rate, growth, and morphology of lettuce under different fractions of red, blue, and green light from light-emitting diodes (LEDs). *Hortic. Environ. Biotechnol.* **2016**, *57*, 573–579. [CrossRef]
16. Izzo, L.G.; Mele, B.H.; Vitale, L.; Vitale, E.; Arena, C. The role of monochromatic red and blue light in tomato early photomorphogenesis and photosynthetic traits. *Environ. Exp. Bot.* **2020**, *179*, 104195. [CrossRef]
17. Son, K.H.; Lee, J.H.; Oh, Y.; Kim, D.; Oh, M.M.; In, B.C. Growth and bioactive compound synthesis in cultivated lettuce subject to light-quality changes. *HortScience* **2017**, *52*, 584–591. [CrossRef]
18. Son, K.H.; Jeon, Y.M.; Oh, M.M. Application of supplementary white and pulsed light-emitting diodes to lettuce grown in a plant factory with artificial lighting. *Hortic. Environ. Biotechnol.* **2016**, *57*, 560–572. [CrossRef]
19. Di, Q.; Li, J.; Du, Y.; Wei, M.; Shi, Q.; Li, Y.; Yang, F. Combination of red and blue lights improved the growth and development of eggplant (*Solanum melongena* L.) seedlings by regulating photosynthesis. *J. Plant Growth Regul.* **2021**, *40*, 1477–1492. [CrossRef]
20. Kim, H.H.; Goins, G.D.; Wheeler, R.M.; Sager, J.C. Green-light supplementation for enhanced lettuce growth under red-and blue-light-emitting diodes. *HortScience* **2004**, *39*, 1617–1622. [CrossRef]
21. Kaiser, E.; Weerheim, K.; Schipper, R.; Dieleman, J.A. Partial replacement of red and blue by green light increases biomass and yield in tomato. *Sci. Hortic.* **2019**, *249*, 271–279. [CrossRef]
22. Han, S.J. Effect of Irradiation of Various Light Sources Including QD(Quantum Dot)-LED on the Growth and Quality of Romaine Lettuce. Master's Thesis, Kangwon National University, Chuncheon, Republic of Korea, 2019.
23. Zhen, S.; van Iersel, M.W. Far-red light is needed for efficient photochemistry and photosynthesis. *J. Plant Physiol.* **2017**, *209*, 115–122. [CrossRef] [PubMed]
24. Zhang, T.; Maruhnich, S.A.; Folta, K.M. Green light induces shade avoidance symptoms. *Plant Physiol.* **2011**, *157*, 1528–1536. [CrossRef]
25. Zou, J.; Zhang, Y.; Zhang, Y.; Bian, Z.; Fanourakis, D.; Yang, Q.; Li, T. Morphological and physiological properties of indoor cultivated lettuce in response to additional far-red light. *Sci. Hortic.* **2019**, *257*, 108725. [CrossRef]
26. Hwang, H.; An, S.; Lee, B.; Chun, C. Improvement of growth and morphology of vegetable seedlings with supplemental far-red enriched led lights in a plant factory. *Horticulturae* **2020**, *6*, 109. [CrossRef]
27. Tan, T.; Li, S.; Fan, Y.; Wang, Z.; Raza, M.A.; Shafiq, I.; Wang, B.; Wu, X.; Yong, T.; Wang, X.; et al. Far-red light: A regulator of plant morphology and photosynthetic capacity. *Crop J.* **2022**, *10*, 300–309. [CrossRef]
28. Shimizu, H.; Saito, Y.; Nakashima, H.; Miyasaka, J.; Ohdoi, K. Light environment optimization for lettuce growth in plant factory. *IFAC Proc.* **2011**, *44*, 605–609. [CrossRef]
29. Saito, Y.; Shimizu, H.; Nakashima, H.; Miyasaka, J.; Ohdoi, K. The effect of light quality on growth of lettuce. *IFAC Proc.* **2010**, *43*, 294–298. [CrossRef]

30. Chen, X.L.; Li, Y.L.; Wang, L.C.; Guo, W.Z. Red and blue wavelengths affect the morphology, energy use efficiency and nutritional content of lettuce (*Lactuca sativa* L.). *Sci. Rep.* **2021**, *11*, 8374. [CrossRef]
31. Kozai, T. *Smart Plant Factory*; Springer: Berlin/Heidelberg, Germany, 2018; pp. 57–209.
32. Kim, H.H.; Goins, G.D.; Wheeler, R.M.; Sager, J.C. Stomatal conductance of lettuce grown under or exposed to different light qualities. *Ann. Bot.* **2004**, *94*, 691–697. [CrossRef]
33. Johkan, M.; Shoji, K.; Goto, F.; Hahida, S.N.; Yoshihara, T. Effect of green light wavelength and intensity on photomorphogenesis and photosynthesis in *Lactuca sativa*. *Environ. Exp. Bot.* **2012**, *75*, 128–133. [CrossRef]
34. Gazula, A.; Kleinhenz, M.D.; Scheerens, J.C.; Ling, P.P. Anthocyanin levels in nine lettuce (*Lactuca sativa*) cultivars: Influence of planting date and relations among analytic, instrumented, and visual assessments of color. *HortScience* **2007**, *42*, 232–238. [CrossRef]
35. Johkan, M.; Shoji, K.; Goto, F.; Hashida, S.N.; Yoshihara, T. Blue light-emitting diode light irradiation of seedlings improves seedling quality and growth after transplanting in red leaf lettuce. *HortScience* **2010**, *45*, 1809–1814. [CrossRef]
36. Bouly, J.P.; Schleicher, E.; Dionisio-Sese, M.; Vandenbussche, F.; Van Der Straeten, D.; Bakrim, N.; Meier, S.; Batschauer, A.; Galland, P.; Bittl, R.; et al. Cryptochrome blue light photoreceptors are activated through interconversion of flavin redox states. *J. Biol. Chem.* **2007**, *282*, 9383–9391. [CrossRef] [PubMed]
37. Nagano, S.; Mori, N.; Tomari, Y.; Mitsugi, N.; Deguchi, A.; Kashima, M.; Tezuka, A.; Nagano, A.J.; Usami, H.; Tanabata, T.; et al. Effect of differences in light source environment on transcriptome of leaf lettuce (*Lactuca sativa* L.) to optimize cultivation conditions. *PLoS ONE* **2022**, *17*, e0265994. [CrossRef]
38. Sasidharan, R.; Chinnappa, C.C.; Voesenek, L.A.C.J.; Pierik, R. The regulation of cell wall extensibility during shade avoidance: A study using two contrasting ecotypes of *Stellaria longipes*. *Plant Physiol.* **2008**, *148*, 1557. [CrossRef]
39. Fan, X.X.; Zang, J.; Xu, Z.; Guo, S.R.; Jiao, X.L.; Liu, X.Y.; Gao, Y. Effects of different light quality on growth, chlorophyll concentration and chlorophyll biosynthesis precursors of non-heading Chinese cabbage (*Brassica campestris* L.). *Acta Physiol. Plant.* **2013**, *35*, 2721–2726. [CrossRef]
40. Spalholz, H.; Perkins-Veazie, P.; Hernández, R. Impact of sun-simulated white light and varied blue: Red spectrums on the growth, morphology, development, and phytochemical content of green-and red-leaf lettuce at different growth stages. *Sci. Hortic.* **2020**, *264*, 109195. [CrossRef]
41. Zheng, L.; He, H.; Weitang, S. Application of light-emitting diodes and the effect of light quality on horticultural crops: A review. *HortScience* **2019**, *54*, 1656–1661. [CrossRef]
42. Folta, K.M.; Maruhnich, S.A. Green light: A signal to slow down or stop. *J. Exp. Bot.* **2007**, *58*, 3099–3111. [CrossRef]
43. Alsiņa, I.; Dūma, M.; Dubova, L.; Šenberga, A.; Daģis, S. Comparison of different chlorophylls determination methods for leafy vegetables. *Agron. Res.* **2016**, *14*, 309–316.
44. Rihan, H.Z.; Aljafer, N.; Jbara, M.; McCallum, L.; Lengger, S.; Fuller, M.P. The impact of LED lighting spectra in a plant factory on the growth, physiological traits and essential oil content of lemon balm (*Melissa officinalis*). *Plants* **2022**, *11*, 342. [CrossRef]
45. Ahmad, M.; Lin, C.; Cashmore, A.R. Mutations throughout an Arabidopsis blue-light photoreceptor impair blue-light-responsive anthocyanin accumulation and inhibition of hypocotyl elongation. *Plant J.* **1995**, *8*, 653–658. [CrossRef] [PubMed]
46. Fuller, K.K.; Dunlap, J.C.; Loros, J.J. Fungal light sensing at the bench and beyond. *Adv. Genet.* **2016**, *96*, 1–51. [PubMed]
47. Bartoli, C.G.; Yu, J.; Gomez, F.; Fernández, L.; McIntosh, L.; Foyer, C.H. Inter-relationships between light and respiration in the control of ascorbic acid synthesis and accumulation in Arabidopsis thaliana leaves. *J. Exp. Bot.* **2006**, *57*, 1621–1631. [CrossRef] [PubMed]
48. Yabuta, Y.; Mieda, T.; Rapolu, M.; Nakamura, A.; Motoki, T.; Maruta, T.; Yoshimura, K.; Ishikawa, T.; Shigeoka, S. Light regulation of ascorbate biosynthesis is dependent on the photosynthetic electron transport chain but independent of sugars in Arabidopsis. *J. Exp. Bot.* **2007**, *58*, 2661–2671. [CrossRef] [PubMed]
49. Chen, W.; Xu, Z.; Liu, X.; Yang, Y.; Wang, Z.; Song, F. Effect of led light source on the growth and quality of different lettuce varieties. *Acta Bot. Boreali-Occident. Sin.* **2011**, *31*, 1434–1440.
50. Yan, Z.; He, D.; Niu, G.; Zhou, Q.; Qu, Y. Growth, nutritional quality, and energy use efficiency in two lettuce cultivars as influenced by white plus red versus red plus blue LEDs. *Int. J. Agric. Biol. Eng.* **2020**, *13*, 33–40. [CrossRef]
51. Chen, X.; Xue, X.; Guo, W.; Wang, L.; Qiao, X. Growth and nutritional properties of lettuce affected by mixed irradiation of white and supplemental light provided by light-emitting diode. *Sci. Hortic.* **2016**, *200*, 111–118. [CrossRef]
52. Matysiak, B.; Ropelewska, E.; Wrzodak, A.; Kowalski, A.; Kaniszewski, S. Yield and quality of romaine lettuce at different daily light integral in an indoor controlled environment. *Agronomy* **2022**, *12*, 1026. [CrossRef]
53. Gude, K.; Talavera, M.; Sasse, A.M.; Rivard, C.L.; Pliakoni, E. Effect of Light Characteristics on the Sensory Properties of Red Lettuce (*Lactuca sativa*). *Foods* **2021**, *10*, 2660. [CrossRef]
54. Meng, Q.; Boldt, J.; Runkle, E.S. Blue radiation interacts with green radiation to influence growth and predominantly controls quality attributes of lettuce. *J. Am. Soc. Hortic. Sci.* **2020**, *145*, 75–87. [CrossRef]
55. Nur Syafini, G.; Abid, A.; Nazrul Hisham, D.; Lip, J. Effect of LED lighting on phytochemical content of lettuce plants (green coral and red coral) grown in plant factory condition. *Food Res.* **2022**, *6*, 80–85.
56. Michell, K.A.; Isweiri, H.; Newman, S.E.; Bunning, M.; Bellows, L.L.; Dinges, M.M.; Grabos, L.E.; Rao, S.; Foster, M.T.; Heuberger, A.L.; et al. Microgreens: Consumer sensory perception and acceptance of an emerging functional food crop. *J. Food Sci.* **2020**, *85*, 926–935. [CrossRef] [PubMed]

57. Paillart, M.J.M.; van der Vossen, J.M.B.M.; Levin, E.; Lommen, E.; Otma, E.C.; Snels, J.C.M.A.; Woltering, E.J. Bacterial population dynamics and sensorial quality loss in modified atmosphere packed fresh-cut iceberg lettuce. *Postharvest Biol. Technol.* **2017**, *124*, 91–99. [CrossRef]
58. Soares, S.; Kohl, S.; Thalmann, S.; Mateus, N.; Meyerhof, W.; De Freitas, V. Different phenolic compounds activate distinct human bitter taste receptors. *J. Agric. Food Chem.* **2013**, *61*, 1525–1533. [CrossRef] [PubMed]
59. Arvanitoyannis, I.S.; Khah, E.M.; Christakou, E.C.; Bletsos, F.A. Effect of grafting and modified atmosphere packaging on eggplant quality parameters during storage. *Int. J. food. Sci.* **2005**, *40*, 311–322. [CrossRef]
60. *ISO 13299:2016*; Sensory Analysis—Methodology—General Guidance for Establishing a Sensory Profile. ISO: Geneva, Switzerland, 2016.

MDPI
St. Alban-Anlage 66
4052 Basel
Switzerland
www.mdpi.com

Plants Editorial Office
E-mail: plants@mdpi.com
www.mdpi.com/journal/plants

www.ingramcontent.com/pod-product-compliance
Lightning Source LLC
Chambersburg PA
CBHW062034220926
51850CB00010B/2546
* 9 7 8 3 0 3 6 5 8 6 4 1 0 *